COLLECTING FIELD
DATA WITH QGIS
AND MERGIN MAPS

KURT MENKE &
ALEXANDRA BUCHA RÁŠOVÁ

Credits & Copyright

Collecting Field Data with QGIS and Mergin Maps

By Kurt Menke & Alexandra Bucha Rášová

Published by Locate Press Inc.

Direct permission requests to info@locatepress.com or by mail:
Locate Press, B102 5212 - 48 ST. Suite 126
Red Deer, AB, Canada, T4N 7C3

Editor   Tyler Mitchell
Cover Design   Andrej Roman
Interior Design Based on Memoir-LATEX
Publisher Website   http://locatepress.com
Book Website   http://locatepress.com/book/mergin-maps

Version: 48bdde8 (2024-09-19)

Contents

Foreword

Are you out in the fresh air yourself, or do you have hundreds of surveyors in the field? Either way, this book shows how to use leading open-source geographical information systems (GIS) to prepare your projects, use them in the field, and present the results afterwards.

Mergin Maps facilitates field surveys, allowing people in the field to start working with it immediately, without any training or prior GIS education. In my recent presentation to high school students, I gave them a prepared Mergin Maps project, and within a few minutes, they were happily outside capturing data from a nearby park. However, the system is so versatile and general that you need a GIS power user to prepare a well-functioning project in the office. After reading this book, you will become that expert!

Kurt can beautifully present and explain GIS – just watch one of his presentations about QGIS at any FOSS4G event. And Saša is the leading technical writer of the Mergin Maps official documentation. What is great about them is that they are not just theorists – both are connected with archaeology, so they understand what field surveying is about!

Preparing a good Mergin Maps project ultimately involves knowledge of specific QGIS areas, such as setting up background layers, preparing survey vector layers, and making everything look nice and clean on the map for your team. Kurt and Saša guide you from the basics to more complex tasks, so by the end, you will know not only how to set up your project but also gain plenty of useful expert tips related to your field surveys.

Peter Petrík
Partner at Lutra Consulting Ltd.

1. Introduction

QGIS is a free and open source geographic information system that probably does not need to be introduced. Suffice to say that QGIS is the world's leading open-source GIS software.

Mergin Maps is an open source platform integrated with QGIS. With Mergin Maps, you can take your QGIS project to the field, collect data using your mobile device and keep them synchronized with QGIS.

Mergin Maps platform consists of:

- *Mergin Maps server*: seamless synchronization and data storage
- *Mergin Maps mobile app*: designed for efficient field surveys
- *Mergin Maps QGIS plugin*: connects QGIS and Mergin Maps

These components are designed to make collaboration easy and effective. A packaged QGIS project is stored in the cloud and multiple contributors can make changes at the same time using a desktop or a mobile device.

Every Mergin Maps project is stored in a *Workspace*. Every workspace is tied to a subscription that defines the number of users, the number of projects and the overall storage size that is available. Users can have access to multiple workspaces.

In this book you will learn to set up a project via QGIS and use the Mergin Maps server to sync that project between your mobile device and QGIS desktop (figure 1.1, on the following page).

Sounds complicated? Don't worry. This book will guide you through the basics as well as show you some best practices and tips and tricks as we explore together the powerful combination of QGIS and Mergin Maps. Before we get started, let's look at these tools and set up everything needed to take full advantage of their capabilities.

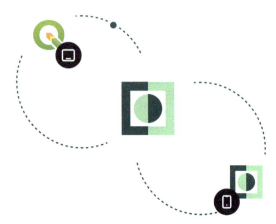

Figure 1.1: The Mergin Maps System (Image credit - Andrej Roman, Graphic designer)

1.1   QGIS installation

In this book, we will be using the current long-term release version of QGIS, version 3.34 Prizren. You can download this version for free from qgis.org.

Follow the download and installation instructions for your platform.

1.2   Signing up to Mergin Maps

To start using Mergin Maps, you need to sign up. You can sign up either through merginmaps.com or from your mobile device using the mobile app.

Here, we will sign up via a web browser.

1. Navigate to merginmaps.com

2. Click the *Start for free* (figure 1.2, on the next page) button

3. Create your account by filling out the form and then clicking on *Sign Up* (figure 1.3, on the facing page). You will receive a confirmation email with a link to verify your email.

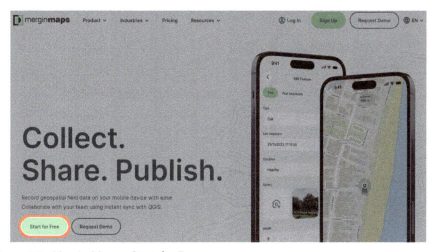

Figure 1.2: Mergin Maps - Start for Free

Let's get started

Username

Email

Password

Confirm password

I accept the Mergin Maps Terms of Service and Privacy Policy.

I want to subscribe to the newsletter

Sign Up

Already have an account? Log in

Figure 1.3: Sign Up for an Account

Check your spam folder if the confirmation email does not appear in your inbox after 5 minutes.

4. Next you will be prompted to create a workspace. Choose an appropriate

workspace name and click on *Create workspace* (figure 1.4).

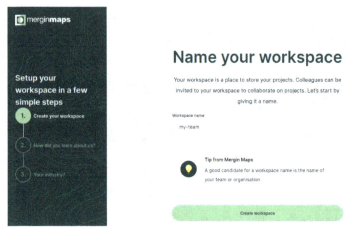

Figure 1.4: Create a Mergin Maps Workspace

> Projects, collaborations and subscriptions in Mergin Maps are tied to your workspace.
> When you create your own workspace, it will automatically start with a free trial.
> To upgrade your subscription after the trial period, check the current subscription
> plans. There are free or discounted plans available for non-profits or academia.

Now you will see the *Mergin Maps dashboard* (figure 1.5, on the facing page). The dashboard is used to manage your projects, workspaces, subscriptions and collaborations. However, for now we will continue with setup. We will get back to the dashboard later in the book.

Remember your Mergin Maps credentials, we will need them to setup the mobile app and QGIS plugin.

1.3   Mergin Maps mobile app installation

Now we will download the mobile app and sign up with our Mergin Maps credentials.

1. Download the Mergin Maps mobile app to your mobile device. The mobile app is available for Android and iOS. You can find it in the app store of your platform.

Figure 1.5: Mergin Maps Dashboard

2. Open the mobile app

3. Tap the account icon in the top right corner of the *Home* screen. Enter your Mergin Maps credentials and *Sign in* (figure 1.6).

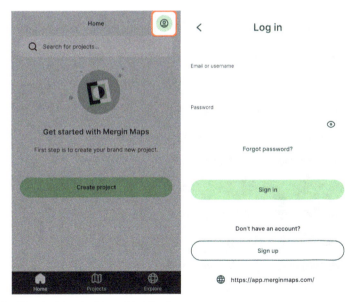

Figure 1.6: Sign in Mergin Maps Mobile App

4. Once logged in, click the back arrow ⟨ < ⟩ to return to the Home screen.

The mobile app is ready to use.

1.4   Mergin Maps QGIS plugin installation

The Mergin Maps QGIS Plugin allows you to synchronize your project to Mergin Maps cloud.

Here are the installation instructions. Make sure you have created an account, and know your username and password.

1. Open QGIS.

2. From the QGIS menu bar choose Plugins | Manage and Install Plugins (figure 1.7).

Figure 1.7: Plugins Menu in QGIS

3. Select the *All* tab, search for *Mergin Maps* and install it (figure 1.8).

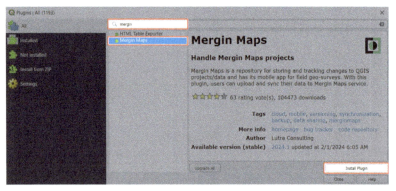

Figure 1.8: Finding and Installing Mergin Maps QGIS Plugin

The Mergin Maps QGIS Plugin has a toolbar (figure 1.9, on the facing page) and appears in the provider section of the *Browser panel*.

Figure 1.9: Mergin Maps Toolbar

4. Click the *Configure Mergin Maps Plugin* button  on the toolbar. You can also right-click on the *Mergin Maps* provider from the *Browser panel* and choose *Configure Mergin Maps Plugin*.

5. Enter your username or email and your password and click *OK* (figure 1.10). Note: you can test your connection before clicking OK, by clicking the *Test Connection* button.

Figure 1.10: Configuring the Mergin Maps QGIS Plugin

You now have all the necessary components installed and configured!

1.5 Typical workflow

The typical workflow for using Mergin Maps consists of the following steps:

- **Preparing project in QGIS:** Load background and survey layers, set up the forms, apply styling to the layers, set up map themes and define the layers to be used for survey purposes.

- **Data/project transfer:** Once the QGIS project is set up, users need to transfer data to the device. This is achieved with the **Mergin Maps QGIS Plugin**.

- **Working with the Mergin Maps mobile app:** Collect and edit field data and synchronize the changes back to the Mergin Maps.

- **Synchronizing the data back to QGIS:** Using the Mergin Maps QGIS plugin the data can be synchronized back to the desktop for management, analysis and cartography.

In the next chapter you will start building a basic QGIS project.

2. Setting up a Project in QGIS

Now that you have everything installed and set up, you will continue by starting a data collection project from scratch in QGIS. This exercise will introduce you to some more advanced project and layer settings which can be helpful in other contexts.

This chapter includes the following tasks:

- Task 1 - Project planning
- Task 2 - Best practices for starting a project
- Task 3 - Adding basemaps
- Task 4 - Creating a survey layer
- Task 5 - Preparing offline basemaps
- Task 6 - Project and Layer level settings
- Task 7 - Exploring the project in Mergin Maps mobile app

After this chapter you will be able to:

- Understand best practices for building a Mergin Maps data collection project in QGIS.
- Add XYZ Tile services to QGIS.
- Create a new layer with specific attributes.
- Generate MB Tile datasets for offline data collection.
- Set up QGIS Map Themes.
- Understand QGIS Project level settings for data collection.
- Create a Mergin Maps project in QGIS and open it in Mergin Maps mobile app.
- Download a public Mergin Maps project in QGIS.

Note that it is possible to set up basic Mergin Maps projects directly via the mobile app, the QGIS plugin, or from the dashboard (app.merginmaps.com). However, the workflow outlined in this first chapter will give you the greatest flexibility in configuring the project.

2.1   Task 1 - Project Planning

Before you get to work building a project in QGIS, it is important to identify your data collection goals and plan your data collection effort. You can then design the project and data structure to meet the needs of your use case.

Here are some initial questions which can help guide your planning effort:

- What kind of data do you need to collect?

- Will the layers be point, line or polygon? You can have a project with multiple survey layers.

- Is there an existing layer(s) you want to use, or will you need to create a new layer(s)?

- What attributes do you need to record (name, description, survey date, etc.) and in what form (string, integer, date, etc.)?

- Will the data be collected by one person or by a team?

In this chapter, you will create a new data collection project for a tree survey. There will be a single survey layer, a point layer representing tree locations. For each tree you will record several attributes including: the tree species, the diameter of the trunk, a photograph, the date and horizontal GPS accuracy.

1. Identify a small study area near your office or home where you will conduct a tree survey. For purposes of learning, this should be a small area, covering just a couple of blocks. Since the author lives in Denmark, the study area shown will be a location in Denmark. Your maps will differ, but the same steps will apply.

2. Determine which coordinate reference system (CRS) you will use for your project. You can simply choose the UTM zone that covers your study area.

2.2 Task 2 - Best practices for building a project

From within QGIS, there are several ways to create a Mergin Maps project. For example, you can open QGIS and from the menu bar click Project | New to start a new QGIS project. Then click the *Create Mergin Maps Project* button [+] from the Mergin Maps toolbar.

You will be presented with three options (figure 2.1). The *New basic QGIS project* option will create a very simple QGIS project in a folder you specify. It will have a survey point layer with predefined attributes and the OpenStreetMap basemap. The *Package current QGIS project* option can be used if you have an existing project you would like to use with Mergin Maps, but one that was not designed with Mergin Maps in mind from the outset. This option will package the project for use in Mergin Maps. In this chapter we will work towards the third option, *Use current QGIS project as is*. For now, you can close this window and move on to the next steps to create your project.

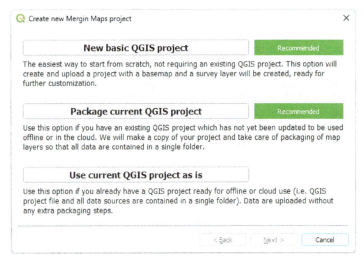

Figure 2.1: Create a New Mergin Maps Project

1. Create a folder for your project.

2. Next save your QGIS project into that folder. From the menu bar choose Project | Save, navigate to your project folder and save the project as tree_survey.qgz.

> This is a very important step. It is crucial that everything related to your QGIS project lies in this folder. This includes any project data files and the QGIS project file!

2.3   Task 3 - Adding basemaps

Basemaps are maps which provide background reference information such as landforms, roads, landmarks and political boundaries. They give context to your study area. For data collection purposes these will be read only layers. Using such a basemap helps reduce the size of the data you are synchronizing from mobile to desktop while simultaneously providing a scalable reference map for your data collection purposes.

Potential sources

There are numerous sources for basemaps for your survey area. Generally, WMS layers or XYZ Tile services work very well for these.

WMS is an acronym which stands for *Web Map Service*. It is an Open Geospatial Consortium (OGC) protocol for serving map data over the internet. The basemap is requested via an HTTP request. Therefore, a network connection is required. The QGIS *Browser panel* has a *WMS/WMTS* data provider. You can right-click on this and choose *New Connection* to add a connection to a WMS service. You simply need the URL of the service.

Yet an easier way of accessing different WMS basemaps is using a plugin. The QuickMapServices plugin is the most popular QGIS plugin for accessing a variety of global basemaps. These are served via WMS requests. You can install this plugin from the Plugin Manager as described in the Introduction to this book. Simply search for QuickMapServices and install it.

There are also regional plugins for country specific basemaps. You will need to inquire if there are any such services in your locale.

The QuickMapServices will appear under the *Web* menu. First you should configure the plugin. From the menu bar choose *Web | QuickMapServices | Settings*. In the *QuickMapServices Settings* window switch to the *More Services* tab. Click the *Get contributed pack* button. You should get a pop up reading *Last version of contrib pack was downloaded*. Click *OK* and *Save*. You will now have access to a large number of global basemaps via the *Web* menu.

XYZ Tile Services provide tiles based on a URL template with values substituted for Zoom Level and the X and Y parameters of each tile. They are (usually) limited to a fixed projection (typically EPSG:3857). QGIS also has a data provider for *XYZ Tile* services which you will also find in the *Browser panel*. QGIS ships with connections to both OpenStreetMap and Mapzen Global Terrain pre-loaded.

Vector Tiles are a great option since they have smaller file sizes, the styles are customizable and they will not pixelate when zoomed in. They can also be easily downloaded to create an offline version. Since QGIS 3.32, there is a *Download vector tiles* algorithm in the *Mergin Maps* section of the *Processing toolbox*. Additionally, this processing algorithm is available via the *Mergin Maps QGIS plugin*. This means that you can access this tool in older QGIS versions (3.16+) if you have *Mergin Maps QGIS plugin* installed. Simply right-click on a Vector tile layer in the *Layers panel* and select *Make available offline* from the context menu (figure 2.2).

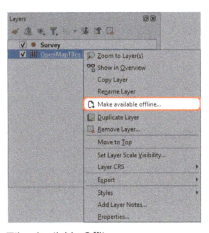

Figure 2.2: Make Vector Tiles Available Offline

Loading Basemaps

1. From the *Browser panel*, find the *XYZ Tiles* provider. Expand it to find the *OpenStreetMap* service. Right-click on *OpenStreetMap* and choose *Add Layer to Project* or simply drag-n-drop it onto the map canvas.

2. Also add the Bing Imagery service by adding a new XYZ Tile service (figure 2.3):

- From the *Browser panel*, right-click on *XYZ Tiles* and select *New Connection* . . .
- For *Name* type *Bing aerial imagery*
- For *URL* type:
    - `http://ecn.t3.tiles.virtualearth.net/tiles/a{q}.jpeg?g=1`
- Press *OK*

Figure 2.3: Setting Up a New XYZ Tile Connection

3. Zoom into the area where you will collect data.

4. Save your project.

2.4   Task 4 - Creating a survey layer

As mentioned at the outset, this project will involve a single point survey layer. However, remember that it is possible to have multiple survey layers, including line and polygon, if required.

It is a best practice to create a new geopackage for each of your survey layers. This will avoid any changes to the database schema. Currently, any change in database schema can cause conflicts during sync. Examples of changing the database schema include: a) Adding or removing a layer to or from an existing geopackage or, b) Adding or removing a field in the attribute table of an existing layer. Generally, if users avoid the above operations, it will be safer for the synchronization process.

It is possible to set up a project for use with Mergin Maps with data stored in a *PostgreSQL* database. There is a tool named *DB Sync* which handles the two-way synchronization between Mergin Maps and PostGIS databases. The synchronization works both ways: changes made in a PostGIS database are automatically pushed to a configured Mergin Maps project and changes made in a GeoPackage in the Mergin Maps project are pushed to the PostGIS database. To configure it, simply navigate to Plugins → Mergin Maps → Configure DB Sync and follow the guide. You can also visit the GitHub repository and read the Quick Start Guide. [a]

---

[a]https://github.com/MerginMaps/db-sync

---

1. From the menu bar again choose Layer | Create Layer | Create Geopackage Layer.

- Create a new *Database* in your project folder.
- Enter a new *Table name* such as tree_survey.
- Choose a *Geometry type* of *Point* and select the appropriate CRS for your area.
- Add the *New Fields* you require for your data collection. Suggestions for a tree survey are below (figure 2.4, on the next page).
- Click *OK* (figure 2.5, on the following page).

Pro tip: Add some extra back-up field attributes when creating a survey layer with different field types (e.g. a couple of string, int, real, date/time fields) and hide them in the form design. These can serve as a backup fields in case you need extra fields later in the survey. At that point you can simply create aliases for these extra fields and add them to the form. You will learn how to set aliases and manage field forms in the next chapter.

---

2. You can symbolize the tree_survey layer as you wish. Below are the steps for making the symbol look like a target using the *Layer Styling Panel*.

Field Name	Type
tree_species	text
tree_type	text
fruit_tree	boolean
circumference	decimal
photograph	text
gps_accuracy	decimal
date	Date & time
notes	text
collector	text

Figure 2.4: Fields to Add to the Survey Layer

Figure 2.5: Creating the Survey Layer

- Make the *Simple marker Fill color* transparent and the *Stroke color* bright red.
- Set the *Size* to 4 and the *Stroke width* to 0.6.

- Click the green plus button to *Add a symbol layer*. Near the bottom of the *Layer Styling Panel* choose the *Cross* icon.

- Set the *Stroke color* to bright red, the *Size* to 8 and the *Stroke width* to 0.6.

> If you choose an SVG marker, be sure to scroll down to the bottom of the *Layer Styling Panel* and find where the file path to the SVG is shown. Click the drop-down menu to the right and choose *Embedded File*. Browse for the SVG you would like to embed, and click Open. It will then read Embedded file.

3. Save your project, which again should be inside your project folder (figure 2.6).

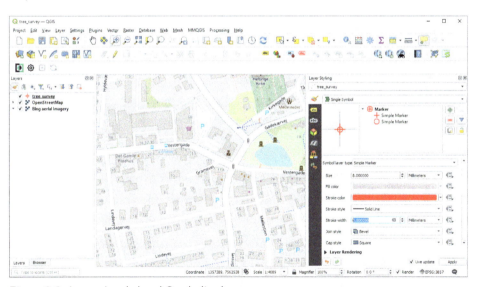

Figure 2.6: Layers Loaded and Symbolized

2.5   Task 5 - Preparing offline basemaps

If you have internet connectivity in the field, you can use a WMS/WMTS or online XYZ tiles as your background map. Keep in mind that they can use up a lot of mobile data during the field survey.

However, if you do not have internet connectivity, you will need to set up offline versions of your background layers. This is done by creating .mbtile files for any of these services we want to access.

1. It is best to set the tile size of your XYZ Tile layer(s) to 256 x 256, so that the labels are readable on mobile devices with high resolution display. To do this, open *Layer properties* for your XYZ Tile layers and switch to the *Source* tab. Set the *Tile Resolution* to *Standard (256 x 256 / 96 DPI)* (figure 2.7).

Tile Resolution	Standard (256x256 / 96 DPI)	▼

Figure 2.7: Setting XYZ Tile Resolution

2. Install the *Zoom Level* plugin. This plugin will display the zoom level of the current map in the status bar which will help when configuring the processing algorithms below. Zoom levels are those used on tiled web maps, typically ranging from 0 (global view) to 20. For more details about zoom levels see the OSM wiki.[1]

3. Turn on just the single OpenStreetMap XYZ Tile layer you want to build MBTiles for. Turn off all other layers.

4. Open the *Processing Toolbox* and search for *MB* to find the *Generate XYZ tiles (MBTiles)* algorithm in the *Raster Tools* toolset (figure 2.8, on the next page).

    a. Set the *Extent* → *Use Current Map Canvas Extent*.

    b. Use the *Zoom Level* plugin to determine the zoom levels you will need. In the example below, a *Minimum zoom* of 18 and a *Maximum zoom* of 20 are being used. This is because when zoomed to the study_area layer, the zoom level reads ~18. *Be careful about including too much detail, otherwise the size of the MBTiles files may become too large.*

    c. Save the *.mbtiles* output file into your project folder.

    d. The file created for this small study area is roughly 4Mb in size. You should try to keep it well below 50Mb for best performance.

---

[1]https://wiki.openstreetmap.org/wiki/Zoom_levels

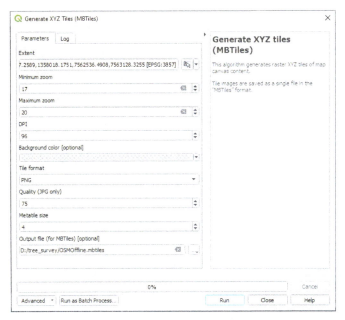

Figure 2.8: Generating XYZ (MBTiles) for the OSM Layer

5. Repeat for the Bing Imagery layer. In this example the MB Tiles were generated only for zoom levels 18 and 19 to keep the resulting file size manageable.

6. Add each MBTiles layer to your project.

7. Select all four basemap layers. To do this, left click on each while holding the Shift key down.

8. Right-click on the four selected layers and choose *Group Selected*. Right-click on the group and *Rename* the group *Baselayers*.

9. Right-click again on the *Baselayers* group and check *Mutually Exclusive Group*. This allows only one base layer to be turned on at a time.

10. Save your project.

Setting up Map Themes

Map Themes allow you to set which layers will participate in the map. Here you will create themes for *OSM Online*, *OSM Offline*, *Bing Imagery Online* and *Bing Imagery Offline*. You will be able to toggle between these map themes when using the Mergin Maps mobile app in the field. This will both allow you to toggle between OSM and Bing imagery basemaps along with offline versions of each, if necessary.

1. To begin, turn on the `tree_survey` layer and one of the MB Tiles layers. Below (figure 2.9) the `OpenStreetMap` MB Tiles layer is being enabled first.

2. From the *Layers panel* click the *Manage Map Themes* button  and choose *Add Theme*. Name it *OSM Offline* (figure 2.9).

Figure 2.9: Adding a Map Theme

3. Repeat this process to create *Map Themes* for *OSM Online*, *Bing Imagery Online*, and *Bing Imagery Offline* (figure 2.10, on the facing page).

2.6   Task 6 - Project and Layer Level Settings

Here you will learn about some useful Project settings. In the project you need to distinguish between:

- **Background layers** that provide context in the field. These will be set up as read only layers. You can use both offline and online layers and they can be raster or vector.
- **Survey layers** into which your field data will be written.

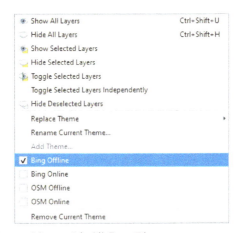

Figure 2.10: Map Themes Menu with All Four Themes

This project involves a single Survey layer: tree_survey. The other layers will be Background layers.

1. From the menu bar choose Project | Properties and switch to the *Data Sources* tab (figure 2.11, on the next page).

2. Here you can control properties of each layer. The *Read-only* column allows you to control which layers are editable. All background layers should be *Read-only*. Therefore, uncheck *Read-only* for any vector layers being used as Survey layers. In this example, the tree_survey layer is the single Survey layer.

The Mergin Maps mobile app also allows you to search attribute forms for data layers in your project.

3. Under the *Searchable* column, uncheck everything other than the Survey layer(s). In this case, only tree_survey is checked as *Searchable* (figure 2.11, on the following page).

Project extent - In the Mergin Maps mobile app, there is an option to zoom to the project extent. If not set, the mobile app will zoom to all visible layers. This is not very useful when you have a layer with a large/global extent (e.g., Open Street Map). Here you will constrain the extent to the current *Map Canvas Extent*.

1. Switch to the *View Settings* tab of *Project Properties*. Click the checkbox

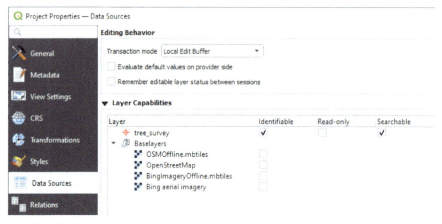

Figure 2.11: Project Properties for Data Sources

in the *Set Project Full Extent* section. Click the button for *Map Canvas Extent* (figure 2.12). Close *Project Properties* when finished.

Figure 2.12: Setting the Project Extent

Mergin Maps Browse Expressions - In the Mergin Maps mobile app you can browse your data. By default, it shows the first column as the header for summary of the field. However, you can define that behavior in QGIS. To change that, you can configure the *Display* properties of the survey layer. Below are instructions for setting up a display expression which will result in: *Apple tree surveyed on 19-03-2022*.

1. Open *Layer Properties* for the tree_survey layer and select the *Display* tab.

2. Click the Expression button to create a display expression.

    a. Expand the *Field and Values* section to reveal the attribute columns for the layer. Double-click on tree_species to add it to the expression.

b. To combine text elements you need to add a *String Concatenation* operator between them. Click on the *String Concatenation* ||| operator.

c. Next you will add some text. Text is always surrounded by single quotes, so type ' surveyed on '. Notice the spaces before and after the phrase.

d. Use a second concatenation operator and double-click on the date column. Below (figure 2.13) the date column has been wrapped in the format_date() function to put the date into a more useful format.

e. Click *OK* to accept the new settings and remember to save your QGIS project.

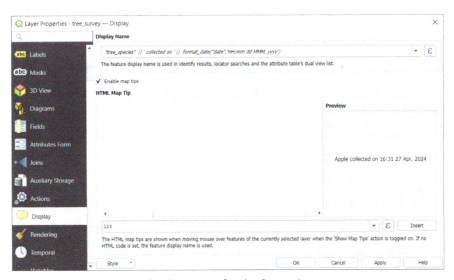

Figure 2.13: Setting Up Display Properties for the Survey Layer

Other Useful Mergin Maps Project Settings

1. Open *Project Properties*.

After installing the Mergin Maps QGIS plugin you will see a *Mergin Maps* tab (figure 2.14, on the next page). This includes sections for *Selective Sync, Photo quality, Snapping, Photo attachments configuration* and *Position tracking. Note: The screen shot includes a sample photo attachments configuration. You will not have this option until you have configured the photo attachment in the next chapter.*

2. Below are brief explanations of each:

    a. *Selective Sync* - this allows you to specify which files from other devices to download during the synchronization process. It is useful when a project contains a lot of data (for example photos) and these data do not necessarily need to be stored on all devices. It can also improve sync time.

    b. *Photo quality* - Use this to specify the quality (*Original, High, Medium, Low*) of your photos. This can be used to reduce the size of photos and save space in your workspace.

    c. *Snapping* - control your field data collection with snapping settings. The options are: *No snapping* - snapping is not enabled (default), *Basic snapping* - features are snapped to the vertices and segments of vector features in the project, or *Follow QGIS project snapping*.

    d. *Photo attachments configuration* - allows you to control how photographs are named using expressions. A common choice is using variables to include layer names, usernames and the date. `@layer_name+'-'+@mergin_username+'-'+format_date(now(),'yyMMddhhmmss')` produces the following photo file name: *username-tree_survey-date.* This can be done individually for each survey layer. This is discussed more thoroughly in the **Advanced Project Configuration** chapter.

    e. *Position tracking* - allows you to record your tracks while collecting field data. This can help you know the extent of areas you have already surveyed in addition to the surveyed features. This is a line feature stored in a standalone geopackage.

3. Enable *Position tracking* with the default *Precision level* (*Normal*). *Precision level* affects the accuracy of the tracking and it may affect the battery usage. For this reason, on longer surveys that don't require high accuracy, you may want to choose the *Low* option. If you need to have more detailed tracking, choose *Best available* accuracy.

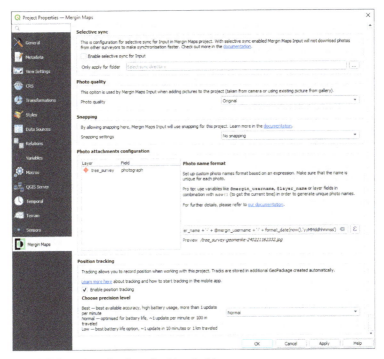

Figure 2.14: QGIS Project Settings for Mergin Maps

4. Enabling tracking means that a new line layer tracking your survey path will be created in your Mergin Maps project called `tracking_layer.gpkg`. This new layer will have 4 attribute columns by default.

- `tracking_start_time` - at date time field with a default value of `@tracking_start_time`
- `tracking_end_time` - another date time field with default value of `@tracking_end_time`
- `total_distance` - a floating point field with a default value of `$length`
- `tracked_by` - a text field with a default value of `@mergin_username`

You can add new fields as needed, however, they should be set up with automatically generated default values since the Mergin Maps mobile app will not open the form for manual inputs. You may use any of the QGIS functions, extra position variables or extra QGIS variables.

5. Click *OK* to accept the settings and dismiss *Project Properties*.

6. Save your project.

Making a Mergin Maps Project

Now that the basic project is set up, you are ready to use the Mergin Maps QGIS plugin to make a Mergin Maps project as described at the beginning of the chapter.

1. Save your project.

2. Make sure you have configured the Mergin Maps QGIS plugin by logging in with your credentials (figure 1.10, on page 7).

3. Click on the *Create Mergin Maps Project* button [+].

4. You will be presented with three options (figure 2.15, on the next page). Since you have set up everything in a single folder, and are using a geopackage for your survey layer, you will use the third option *Use current QGIS project as is*.

5. The *Create new Mergin Maps project* window will appear. Enter a *Project Name* and click *Finish* (figure 2.16, on the facing page).

Note that you have the option to make the project a Public project. Public projects are discoverable by anyone.

6. The *Mergin Maps - Synchronization* window will appear. After it has completed synchronizing, the *Create Project* window will appear telling you the project has uploaded successfully. Click *Close*.

7. Switch to the *Browser panel* and expand the *Mergin Maps* provider. Expand your project entry to see the project folder contents (figure 2.17, on page 28).

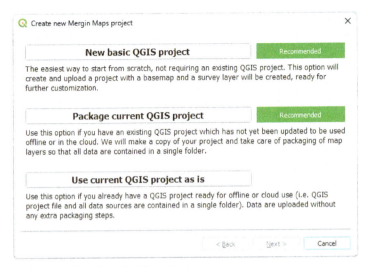

Figure 2.15: Create a New Mergin Maps Project

Figure 2.16: Naming the New Mergin Maps Project

2.7 Downloading a public Mergin Maps project

The Mergin Maps project for this chapter is public and you can download it if you want to check your settings and inspect the solution. It can be downloaded by using this link.[2]

---

[2]https://app.merginmaps.com/projects/qmm-book/qmm-book-tree_survey-chapter1/tree

Figure 2.17: Mergin Maps Project Content in QGIS Browser

You can also download it using the *QGIS Browser Panel*.

1. Right-click on the *Mergin Maps* data provider and select *Explore Public Projects* from the context menu (figure 2.18).

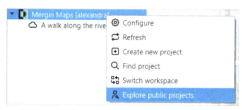

Figure 2.18: Explore Public Mergin Maps Projects from QGIS Browser

2. Search for the project. It is named *qmm-book-tree_survey-chapter1* (figure 2.19, on the next page). Use the *Open project* button to download it to your computer.

3. Create a folder for this project and select that folder in the *Open Directory* window.

4. The project will be downloaded. When finished click *Yes* in the *Project download* window to open the project.

2.8   Task 7 - Exploring the project in Mergin Maps mobile app

Let's look at the project in Mergin Maps mobile app.

1. Open the mobile app. Make sure you are signed in (as described in Mergin

Figure 2.19: Find and Download a Public Mergin Maps Project

Maps mobile app installation, on page 4).

2. Navigate to the *Projects* tab in the bottom navigation panel. Find your project and tap on it to download it (figure 2.20).

Figure 2.20: Downloading a Project in Mergin Maps Mobile App

3. Switch to the *Home* tab (figure 2.21, on the following page). Here you can see all projects that are downloaded to your mobile device. Tap on the project

to open it in the mobile app.

Figure 2.21: Downloaded Projects in the Home Tab

4. The project opens. Move the map by dragging around. Zoom in and out by pinching open or close.

5. There are useful tools in the bottom navigation panel.

- *Sync* can be used to synchronize the project.

- *Add* is used to record features (we will look into this in more detail in the Conducting a Field Survey, on page 43).

- *Layers* display the overview of layers in the project (figure 2.22, on the facing page).

- *More* provides more options (figure 2.23, on the next page).

6. Tap the *More* button. Try the settings you have defined when creating the project in QGIS, such as *Zoom to project* to display the **Project extent** or *Map themes* to switch between the map themes defined in the project (figure 2.24, on page 32). *Position tracking* is included in the menu if you have enabled this option in QGIS.

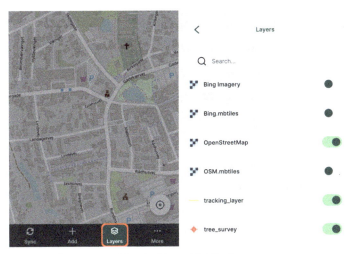

Figure 2.22: Layers Overview in Mergin Maps Mobile App

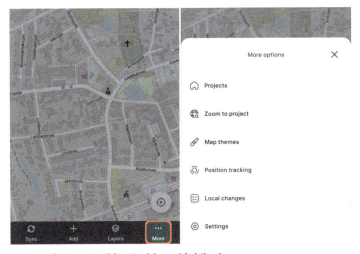

Figure 2.23: More Options in Mergin Maps Mobile App

7. If something does not work as expected, go back to QGIS and make changes in the project settings. Save the project, then use the *Sync* button 🔄 to synchronize changes to the cloud.

8. In the mobile app, tap the *Sync* button to synchronize changes.

Figure 2.24: Map Themes in Mergin Maps Mobile App

Since *qmm-book-tree_survey-chapter1* is a public project, it can be also down-loaded directly to the mobile app. On the main page of the app, tap the *Explore* button and search for this project (figure 2.25). You can download it and compare it with your solution.

Figure 2.25: Explore Public Projects in Mergin Maps Mobile App

3. Configuring Survey Layers

Tables in QGIS can be viewed in the default table view or as a form. QGIS has widgets which allow you to configure the interface for each field in the attribute table form view. The Mergin Maps mobile app honors all these settings. Therefore, you can configure the data collection form via a QGIS attribute table and form widgets. Here you will learn how to configure these for the survey layer created in the last chapter.

This chapter includes the following tasks:

- Task 1 - Configuring Field Widgets
- Task 2 - Using the QGIS Drag and Drop Designer

After this chapter you will be able to:

- Configure QGIS field widgets for data collection.
- Use the QGIS Drag and Drop Designer for field forms.

3.1 Task 1 - Configuring Field Widgets

In this task, you will configure the field form. You will be configuring the form to make data collection more intuitive for the field surveyors. This will involve using Aliases. Aliases allow you to replace field names with longer text strings. These can be formed into short questions to prompt the data collector with clearer instructions for each data element. You can also set default values and use specific widget types to make data entry easier and less prone to error.

1. Open *Layer Properties* for the t ree_survey layer and switch to the *Attributes form* tab.

2. The *Fields* are listed under the *Available widgets* heading. You will select each, set the widget type and configure the settings.

3. Start by selecting the fid field. This field is part of the geopackage format and will auto populate as features are collected. Therefore, the surveyor does not need to see it. Set the *Widget Type* to *Hidden* (figure 3.1).

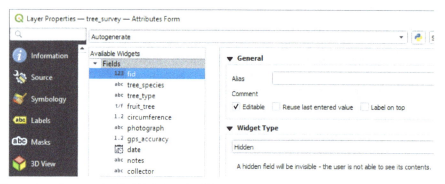

Figure 3.1: Setting the Field Widget for fid to Hidden

As you can see from the list, there is a large array of widget types. For each widget there are additional configuration options. These include default values, conditional visibility and constraint enforcement. Each also includes the option to add an Alias. An Alias allows you to enter a human readable name for fields. It will be displayed in the feature form, the attribute table, or in the Identify results panel. When using Mergin Maps, this gives you an option to pose a question for each field. This can serve as a prompt to the surveyor. For example, the alias for tree_species could be: *What type of tree is it?*

4. Select the tree_species field.

    a. Set the *Widget Type* to *Text Edit*. This widget allows the data collector enter some text using the keyboard. In this case, they can type a species name or common name.

    b. Under *General* for *Alias* type: *Enter the tree species.*

    c. Note: If this is a required field, you can check *Not null* in the *Constraints* section. This option is available for all widgets. If *Not null* is checked, the data collector will get a warning if this is left blank. Constraints are discussed in more detail in Advanced Project Configuration, on page 65.

    d. Note: There is an option to set this up as a *Multiline* field. This is not relevant here but may be for other fields.

    e. Note: If there were a dominant tree type in the study area this could be entered as the *Default Value*.

> NOTE: This field could also be configured using a Value Map widget if there were a specific list of trees to be mapped. For example, Sycamore, Oak, Beech etc. You will see an example of using this widget next.

5. Next, select the tree_type field.

    a. Set the *Widget Type* to *Value Map*.

    b. Under *General* for *Alias* type: *Deciduous or evergreen?*

    c. For both *Value* and *Description* type the following values: *Deciduous, Evergreen* (figure 3.2).

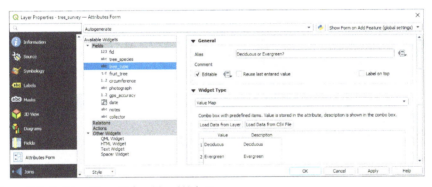

Figure 3.2: Setting Up a Value Map Widget

6. Next, select the fruit_tree field.

    a. Since this is a *Boolean* field, the *Widget Type* is automatically set to *Checkbox*.

    b. Under *General* for *Alias* type: *Fruit tree?*

There are also *Text*, *HTML* and *Spacer* widgets which can be used to design your data collection form. These are covered in more detail in the Advanced Project Configuration chapter.

7. Next, select the `circumference` field.

    a. Since this is a numeric field, set the *Widget Type* to *Range*.

    b. Under *General* for *Alias* type: *Measure the circumference in cm at 1,3 meter above ground.*

    c. There are three interfaces for Range widgets: *Editable*, *Slider* and *Dial*. The last two are more suited to integers. Choose *Editable*. This will provide a numeric keypad to enter the number.

    d. You can configure the *Minimum* and *Maximum* values and the *Step*. This will limit the entered values into the expected range. Here 1000 has been entered for the *Maximum* and 5 for the *Minimum* with a *Step* of 0.5 (figure 3.3). Note: depending on the version of QGIS you are using, you may need to set the *Precision* to 2.

Figure 3.3: Setting Up a Range Widget

8. Next, select the `photograph` field.

    a. Set the *Widget Type* to *Attachment*.

    b. In the *Path* section, set *Store path as* to *Relative to project path*. Note that you can also configure a default value so that the photographs are stored in a dedicated folder. For example, you could create a photos folder in your project folder and use the following expression as the *Default Path* value: @project_home + '/photos'. This and other more advanced photograph specific settings is covered in more detail in the Advanced Project Configuration chapter.

   c. Also in the *Path* section, set the *Storage Mode* to *File Paths*.

   d. Under *Integrated Document Viewer*, set *Type* to *Image*.

> It is possible to configure a Mergin Maps survey layer so that multiple photographs can be attached to each feature. This is covered in more detail in the Advanced Project Configuration chapter.

9. Next, select the gps_accuracy field. Mergin Maps mobile app provides the option to access GPS information using extra position variables (figure 3.5, on page 40). Here you can use the variable @position_horizontal_accuracy as a Default value to collect the accuracy.

   a. Set the *Widget Type* to *Hidden*.

   b. Under *Defaults*, type @position_horizontal_accuracy. This variable will be used by the Mergin Maps mobile app to populate the data from your mobile device. QGIS cannot give you a preview of the value because this is a Mergin Maps variable, not a QGIS variable. However, the Mergin Maps app will be able to populate the data.

10. Next, select the date field. Here you can use a default setting to make data collection easier.

   a. Set the *Widget Type* to *Date/Time*.

   b. In the *Constraints* section check *Not null*.

   c. Under *Defaults*, for *Default value* type now(). This function will default to the current date and time.

> Note: Under *Defaults*, if you check *Apply default value on update* the date will be updated every time this feature is edited, whether in the Mergin Maps app or QGIS.

11. Next, select the `notes` field.

   a. Set the *Widget Type* to *Text Edit*.
   b. Under *General* for *Alias* type: *Enter any notes*.
   c. Check *Multiline* so that longer comments can be entered.

12. Lastly, select the `collector` field.

   a. Set the *Widget Type* to *Text Edit*.
   b. Under *Defaults* you can either type your name surrounded by single quotes. For example, 'Otto Maddox' or you can use the variable `@mergin_username`. This equals the username you created with your Mergin Maps account. In either case the *Preview* will display the value.

Additional Variables

The `@mergin_username` variable was just mentioned as an option for the default value in a field such as the `collector`. This variable is added when the Mergin Maps QGIS plugin is installed. There are several additional Mergin Maps QGIS plugin variables you can use as default values (figure 3.4, on the next page).[3]

As you learned when configuring `gps_accuracy` field, the Mergin Maps mobile app also provides the option to access GPS information using extra position variables (figure 3.5, on page 40). These can be used as default values in feature forms. For example, you could have another column for altitude, and populate the default value with `@position_altitude`. Importantly, QGIS will not recognize these. However, the Mergin Maps mobile app will when you are collecting data.[4]

[3]https://merginmaps.com/docs/layer/plugin-variables/
[4]https://merginmaps.com/docs/layer/position_variables/

Mergin Variable Name	Scope	Description
@mergin_username	Global	Name of the user currently logged in to Mergin Maps
@mergin_user_email	Global	Email of the user currently logged in to Mergin Maps
@mergin_url	Global	URL of the Mergin service
@mergin_project_name	Project	Name of the active Mergin project
@mergin_project_owner	Project	Name of the owner of the active Mergin project
@mergin_project_full_name	Project	Owner and project name joined with a forward slash
@mergin_project_version	Project	Current version of the active Mergin project

Figure 3.4: Mergin Maps QGIS Plugin Variables

Note that location permission has to be allowed and location service has to be enabled on your mobile device.

3.2 Task 2 - Using the Drag and Drop Designer

By default, every QGIS table has a Form view. These forms are automatically generated and contain all the fields in the layer. The form is used to collect data in the Mergin Map app. However, you might want to change the order of the fields. Also, there may be some fields that do not need to be displayed during the survey. This might include the fid column or fields with default values using expressions and variables such as date and gps_accuracy. The Drag and Drop Designer allows you to create an attribute form which can even include different tabs (containers) to better present the attribute fields. This can be especially useful if you have a long survey form.

1. Continue from the last section.

2. From the drop-down menu where it reads *Autogenerate* select *Drag and Drop Designer*.

Multiple containers can be added to the attributes form to group related fields together. You can then position different fields into the newly generated containers.

3. Click the green plus button (Add a new tab or group to the form layout) to

Position Variable Name	Type
@position_coordinate	A point with the coordinates in WGS84.
@position_latitude	Latitude.
@position_longitude	Longitude.
@position_altitude	Altitude.
@position_direction	The bearing measured in degrees clockwise from true north to the direction of travel.
@position_ground_speed	The ground speed, in meters/sec.
@position_vertical_speed	The vertical speed, in meters/sec.
@position_magnetic_variation	Magnetic declination.
@position_horizontal_accuracy	The accuracy of the provided latitude-longitude value, in meters.
@position_vertical_accuracy	The accuracy of the provided altitude value, in meters.
@position_from_gps	True, if recorder/edited feature's geometry correspond with current user's position.
@position_satellites_visible	Number of visible satellites.
@position_satellites_used	Number of satellites used to calculate the position.
@position_gps_fix	GPS fix, e.g. "RTK float".
@position_gps_antenna_height	Antenna height as defined in GPS settings.
@position_provider_type	GPS device type.
@position_provider_name	GPS device name.
@position_provider_address	GPS device address.
@position_hdop	Horizontal dilution of precision (HDOP).
@position_vdop	Vertical dilution of precision (VDOP).
@position_pdop	Position (3D) dilution of precision (PDOP).

Figure 3.5: Extra Position Variables

create a new tab and name it *Tree Info* (figure 3.6, on the next page).

4. Select the fid field and click the minus button to remove it from the form's layout. It will be auto-populated and the survey does not need to see it. This will make the survey form cleaner.

5. Drag the fields tree_species, tree_type, fruit_tree, circumference and photograph beneath the *Tree Info* container.

6. Add another container named *Other details*.

7. Drag the fields date, notes and collector beneath the *Other details* con-

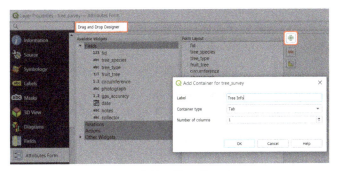

Figure 3.6: Drag and Drop Designer - Add a New Container

tainer.

8. Click *OK* to accept these changes and dismiss the *Layer Properties* window. Your final form design should look similar to the one on figure 3.7.

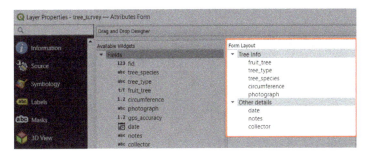

Figure 3.7: Attributes Form with Attributes Organized into Two Containers

9. Save your project.

10. You have completed the basic project configuration in QGIS and your project is ready to take into the field! Since you have made changes to your project you will synchronize it. Make sure you are logged in with your credentials via the Mergin Maps toolbar (as described in Mergin Maps QGIS plugin installation, on page 6). Click the *Synchronise Mergin Maps Project* ⟳ button in Mergin Maps toolbar. This uploads your new changes to the Mergin Maps cloud.

The Mergin Maps project for this chapter is public and you can download it if you want to check your settings and inspect the solution. It is named *qmm-*

book-tree_survey-chapter2.

Download it by following the steps in the task Exploring the project in Mergin Maps mobile app, on page 28 in the previous chapter or by clicking this link.[5]

In this chapter you configured a point survey layer. The same steps can be used to configure the attribute form for line or polygon survey layers. In the next chapter you will learn how to take this project outside on your mobile device and conduct the tree survey.

---

[5]https://app.merginmaps.com/projects/qmm-book/qmm-book-tree-survey-chapter2/tree

4. Conducting a Field Survey

Now that the QGIS project is ready, we can take it to the field and collect some data.

This chapter includes the following tasks:

- Task 1 - Collecting Field Data with Mergin Maps mobile app
- Task 2 - Enable Position Tracking
- Task 3 - Navigation
- Task 4 - Synchronizing the Data Back to QGIS
- Task 5 - Using Mergin Maps Processing Algorithms
- Task 6 - Collecting Data with Multiple Survey Layers

After this chapter you will be able to:

- Download QGIS projects to your mobile device.
- Use the Mergin Maps mobile app for collecting field data.
- Understand how to enable and use position tracking on the mobile device.
- Navigate to features.
- Use the Mergin Maps QGIS plugin to synchronize the data back to QGIS.
- Use Mergin Maps Processing Algorithms to track changes in a project.
- Collect data with multiple survey layers.

4.1   Task 1 - Collecting Field Data with Mergin Maps Mobile App

For this task ensure that you have installed the Mergin Maps mobile app on your mobile device and that you are signed in (as described in Mergin Maps mobile app installation, on page 4).

1. Open the Mergin Maps mobile app on your mobile device.

2. On the *Projects* tab, find your tree survey project. Click on the download icon ⬇ next to it to download it.

3. Tap on the project to open it.

4. The map will appear as you saved it in QGIS (figure 4.1, on the facing page). An oval GPS accuracy indicator will be at the bottom left corner of the map, giving you an indication of your current horizontal GPS accuracy.

> Your GPS accuracy depends on several parameters such as the GPS chip in your device and the sky view while you are collecting data. When recording a point, it is recommended to note the limitations of your GPS accuracy. If your horizontal accuracy is above the GPS accuracy threshold, the color of the GPS accuracy indicator will change to yellow. The default threshold is 10 meters. This value can be adjusted in *Settings*. If you'd like to have better horizontal GPS accuracy, you can wait for your device to acquire a better GPS signal. Alternatively, you can connect your device to an external (Bluetooth) GPS receiver. In this case, you might also need to set up the *GPS antenna height* in the *Settings*.

5. Clicking on the ⦂More button will reveal *Projects, Zoom to project, Map themes, Position tracking* if enabled in the QGIS project, *Local changes* and *Settings* (figure 4.2, on page 46). Try toggling between the different *Map themes* (both online and offline) and explore the *Settings*.

6. To record a tree, click the +Add button. A menu will appear atop the map which allows you to select your survey layer (figure 4.3, on page 47). The arrow to the left ‹ will take you out of data collecting mode. In this case, we only have one survey layer tree_survey.

The cross-hairs symbol ⨁ on the map represents the point that will be cap-

Figure 4.1: Project Loaded in Mergin Maps Mobile App

tured. You can drag the map to change its position. To recenter the map to your current position based on GPS, tap the GPS button ⊙ . To capture the point, click the large green *Record* ⬤ Record button on the bottom menu bar.

7. The feature form will come up. If you set up the Drag and Drop Designer containers, you will see the two tabs at the top of the form (figure 4.4, on page 48). Fill in the information on both tabs.

8. When ready to record the data click the green check mark *Save* ✔ button.

9. After you have collected several trees, you can explore the features you've collected. Click the *Layers* 🗇 button. Click on the *tree_survey* layer. If you set up the expression at the end of the Setting Up a QGIS Project chapter - Task 6, it should resemble figure 4.5, on page 49.

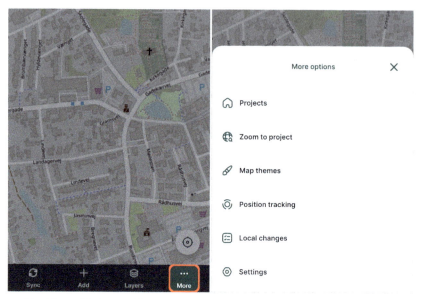

Figure 4.2: Menu Choices in Mergin Maps Mobile App

10. After you a finished with the field session, tap on the *Synchronize* button. Your changes will be pushed up to Mergin Maps.

> It is also possible to expedite survey work by setting Mergin Maps to *Reuse the last entered value*. To configure this option, click the *More* button and choose *Settings*. Find the *Recording* section. Toggle on *Reuse last entered value*. Click the Back button to return to the survey. In the form you will see checkboxes next to each attribute. Check the attributes, whose values you want to reuse in features you will create next. When creating another feature, the checked attributes in the form will contain the values from the previous feature.

4.2 Task 2 - Enable Position Tracking

In this task, you will use position tracking, which was enabled in QGIS *Project Properties* during the initial project set up (see Task 6 - Project and Layer Level Settings, on page 20.)

Figure 4.3: Selecting Active Survey Layer

1. Head outside to your survey area and open the Tree survey project that you used in Task 1 in the mobile app.

2. Click on the [More] button and from the context menu choose *Position tracking* (figure 4.6, on page 49).

3. The *Position tracking* window will open. Click *Start tracking* to begin (figure 4.7, on page 50).

4. A small green oval tracker icon will appear on the left bottom corner of the map. As you travel to your first feature of this survey, your path will be automatically digitized onto the map. As you continue to move, the small tracker icon ⊙ 00:00:32 will display the time you have been moving.

5. When you wish to stop tracking click that small oval tracker icon to open the *Position tracking* window. Click *Stop tracking* (figure 4.8, on page 51). Notice that a summary of the time and distance you have travelled is shown!

Figure 4.4: Tree Survey Form

When you sync the project and open it in QGIS, you will have a tracking layer along with the survey layers.

4.3 Task 3 - Navigation

Points in your survey layers can be staked out. This means that you can use the Mergin Maps mobile app to navigate towards the selected point. The app will show you both the direction and distance to your target as you navigate. This can be helpful when surveyors need to find points for a follow up data collection effort.

1. The units for your navigation can be set in your QGIS *Project Settings*. On the *General* tab, find the *Measurements* section. Set the *Units for distance measurement* accordingly (figure 4.9, on page 51). Click *OK* to accept. Remember

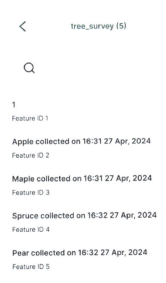

Figure 4.5: Browsing Features in Mergin Maps Mobile App

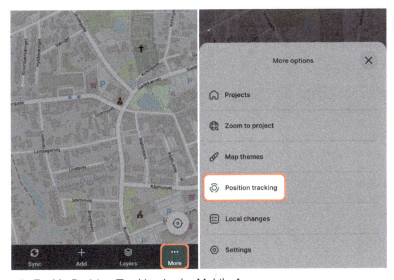

Figure 4.6: Enable Position Tracking in the Mobile App

to *Save* your project and *Synchronize* it using the Mergin Maps toolbar.

2. Now that your units are configured, head out for your survey. Remember to

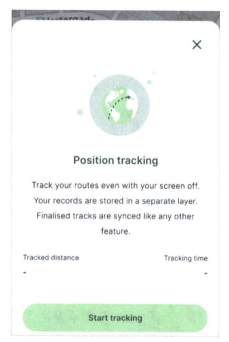

Figure 4.7: Start Position Tracking in the Mobile App

Synchronize the project also on your mobile device so that any changes to your measurements units will be in place.

3. Select a point by clicking on it. The point will be highlighted and you have the choice of Editing or Staking out. Choose *Stake out* (figure 4.10, on page 52).

4. The direction will be shown on the map relative to your current position. A dark green straight-line between you and the destination will appear. The distance will also be displayed in the units you specified. As you move towards the point, the distance will update (figure 4.11, on page 52).

5. As you get within 1 meter of the point, Mergin Maps switches to short navigation mode (figure 4.12, on page 53). Precise stake out of the point (distance less than 10 cm) will be highlighted in green.

6. Remember to *Synchronize* 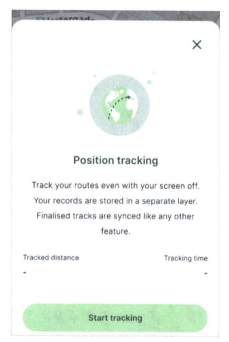 your changes if you have collected additional data or edited data.

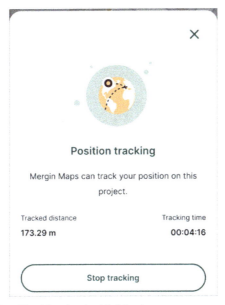

Figure 4.8: Stop Position Tracking in the Mobile App

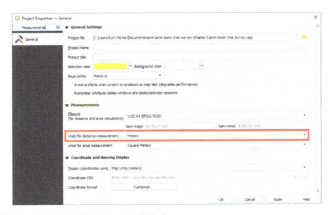

Figure 4.9: Stake Out Unit Settings in QGIS

There is an *Automatically sync changes* setting in the Mergin Maps mobile app *Settings*. With this option enabled, each time you save changes, the app will sync automatically. This of course, depends on having network connectivity.

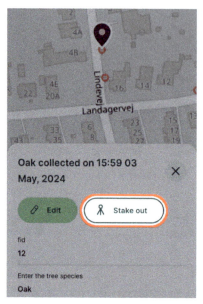

Figure 4.10: Start Staking Out in the Mergin Maps Mobile App

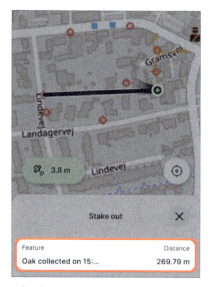

Figure 4.11: Navigating to the Point in the Mergin Maps Mobile App

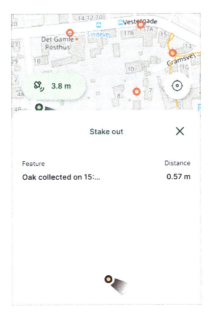

Figure 4.12: Short Navigation Mode in the Mergin Maps Mobile App

4.4 Task 4 - Synchronizing the Data Back to QGIS

Here you will see how easily you can pull the data you collected back into QGIS.

1. Open QGIS with the Tree survey project open.

2. On the Mergin Maps toolbar, click the *Synchronise Mergin Maps Project* button. A *Project Status* window will open detailing the changes (figure 4.13, on the next page). Click the yellow *Sync* button.

3. The project will synchronize, and the tree points you collected will appear on the map canvas (figure 4.14, on the following page).

Note: If you enabled position tracking for all or part of your survey, there will be a line tracking layer and your track will be shown.

You can repeat the steps you have learned to continue your data collection effort over a period of time.

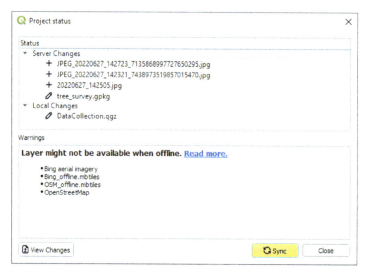

Figure 4.13: Mergin Maps Project Status

Figure 4.14: Tree Locations Synchronized Back to QGIS along with a Tracking Layer

4.5 Task 5 - Using Mergin Maps Processing Algorithms

After you have collected data from your survey, you can analyze the results with Mergin Maps processing algorithms. When you install the Mergin Maps

QGIS plugin a set of Mergin Maps processing algorithms is added to the QGIS Processing toolbox. These include *Create diff*, *Create report*, *Download vector tiles* and *Extract local changes* (figure 4.15). In this task you will see examples of how to use these tools and what they produce.

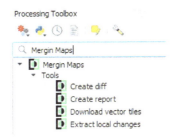

Figure 4.15: Mergin Maps Processing Algorithms

Create diff

1. After you have multiple survey sessions and/or some editing that has been done in QGIS, you will have multiple versions of your project. Mergin Maps keeps track the version history. Login to your dashboard (app.merginmaps.com) and open your tree project. Switch to the *History* tab to see the different versions of your project (figure 4.15). The version number is shown in the first column.

Files	Map	History	Collaborators	Settings				⤓ Download	⟳ Clone

Version	Created	Author	Files added	Files edited	Files removed	Size	
v16	1 minute ago	geomenke	0	2	0	9.48 MB	⤓
v15	17 hours ago	geomenke	0	1	0	9.48 MB	⤓

Figure 4.16: Mergin Maps Dashboard Project Version History

2. Return to QGIS and open the *Processing toolbox*. Find the *Mergin Maps* toolset and open the *Create diff* algorithm (figure 4.17, on the next page).

    a. *Project directory* - The folder containing your Mergin Maps project.

   b. *Input layer* - The survey layer of your choice.

   c. *Start version* - The start version of the difference calculation.

   d. *End version* - The end version of the difference calculation.

   e. *Diff layer* - Save this to a new geopackage in your project folder.

   f. *Run* the tool.

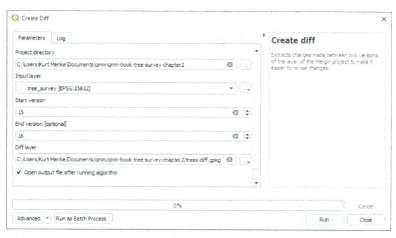

Figure 4.17: Create Diff Algorithm Ready to Run

3. A new layer is added to QGIS showing which features were *Inserted*, *Deleted* or *Updated* with new information between the two versions of the project. If you right-click on the layer and choose *Show Feature Count* you can quickly see how many features fall into each category (figure 4.18, on the facing page).

Create report

You can also generate a tabular report detailing changes across multiple versions of a project. The output is a CSV file, which can be opened in QGIS along with common text and spreadsheets programs.

1. Open the *Create report tool* (figure 4.19, on the next page).

   a. *Project directory* - The folder containing your Mergin Maps project.

   b. *Start version* - The start version for the report.

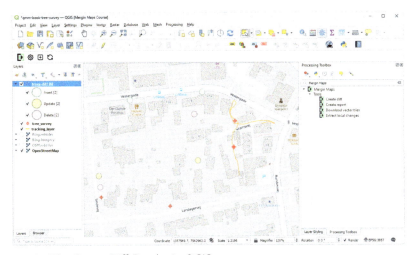

Figure 4.18: The Create Diff Results in QGIS

    c. *End version* - The end version for the report.

    e. *Report* - The output CSV file saved into the project folder.

    f. *Run* the tool.

Figure 4.19: The Create Diff Results in QGIS

2. The report is added to QGIS as a table. Right-click on it and select *Open Attribute Table* to open it. It contains a row for each version in the range you

chose detailing changes for each version (figure 4.20). If you have enabled *Positional tracking* in one of those project versions, the length will be included in the length field .

Figure 4.20: The Mergin Maps Report

Extract local changes

Local changes of a specific layer can also be extracted using the *Extract local changes* tool in the Processing toolbox. Local changes are those done in QGIS (figure 4.21).

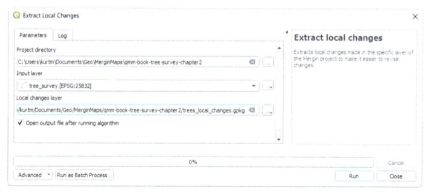

Figure 4.21: Extract Local Changes Algorithm

4.6 Task 6 - Collecting Data with Multiple Survey Layers

In this final task you will use a slightly modified project to collect point, line and polygon features. A public project has been created for this task. You will clone it and use it in the field.

1. It is named qmm-book--neighborhood-survey-chapter3. In your web browser, go to app.merginmaps.com and log into your account.

2. On the *Projects* tab, find the *Browse community projects* button (figure 4.22).

Figure 4.22: Browse Public Projects on Mergin Maps Dashboard

3. Search for qmm to find projects associated with this book.

4. Find qmm-book--neighborhood-survey-chapter3 and click on it to open the project (figure 4.23).

Figure 4.23: The Chapter 3 Project Opened on Mergin Maps Dashboard

5. Click on the *Clone* button. Give the project a name and select your *Target workspace* (figure 4.24). This is the workspace where the project will be created.

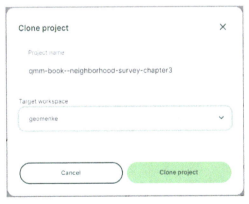

Figure 4.24: Cloning a Mergin Maps Project

6. Switch to your mobile device. If you have the previous map open click on the **More** button and choose *Projects* from the menu (figure 4.25).

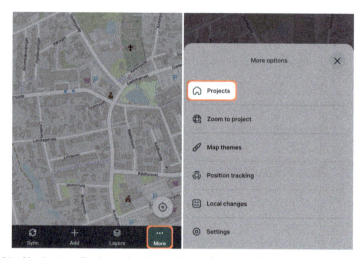

Figure 4.25: Navigating Back to the Main Page of the Mobile App

7. Find the cloned project in the list. Click the download button ⬇ next to it to download it to your mobile device.

8. Once downloaded, click on the project to open it. We can now head outside to collect some tree, path and park data.

9. When you are ready to collect your first feature, click *Add* ⊞ button. The layer selector at the top of the map will display the active layer. To change this, click the layer selector at the top to open the *Choose Active Layer* menu. Select the layer you want to use, e.g. paths (figure 4.26, on the facing page).

10. There are two methods of capturing lines and polygons: adding vertices one by one or using the streaming mode to capture features based on your position.

- To add vertices one at a time, use the *Add* ⊞ and *Remove* ▬ buttons to enter and delete vertices until you have digitized the feature. Click the *Record* ⊙ button to open the form and enter the data. Click the *Save* ✓ button when finished.

- To begin using *Streaming* mode, click the ⟨ ⚐ ⟩ button. The

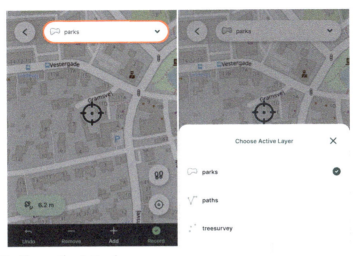

Figure 4.26: Choose the Active Layer

Streaming mode window opens.

– Click *Start streaming mode* button. The *Active Streaming* mode icon ⚏ streaming will appear.

– When you are finished click *Active Streaming* mode icon again and click *Stop streaming mode*.

– Click the *Record* ✅ button to open the form and enter the data. Click the *Save* ✅ button when finished.

You can configure the *Streaming mode* settings. Simply click on the ⚬⚬⚬ More button, choose *Settings* and find the *Streaming mode* section. You have choices for *Interval threshold type* (*Time elapsed* or *Distance traveled*) and *Threshold interval* (figure 4.27, on the next page).

11. Attributes and geometry of recorded features can be changed. Recorded features can be deleted. Tap on a tree, path or park that you have surveyed. The *Edit* button 🖉 Edit opens the form where you can modify the attribute values. If you want to delete the feature, use the *Delete* button 🗑 Delete.

Streaming mode

Interval threshold type
Choose a type of threshold for   Time elapsed
streaming mode

Threshold interval
Streaming mode will add a point   3 s
to the object at each interval

Figure 4.27: Streaming Mode Settings

Tap the *Edit geometry* button 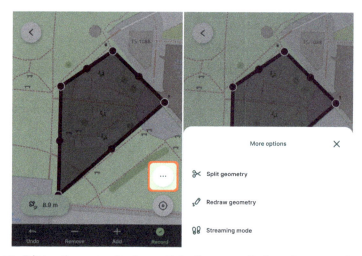 to change the geometry by adding, removing or moving the vertices. It is also possible to *Redraw geometry* or *Split* it into multiple features (figure 4.28).

Figure 4.28: Editing Geometry Options - Split Geometry, Redraw Geometry, Streaming Mode

12. Synchronize the project when you are finished with your survey.

13. Open QGIS, open the *Browser Panel,* and find the cloned project under the *Mergin Maps* data provider. It will have a small cloud icon ☁ next to it.

14. Right-click on the project and choose *Download* from the context menu (figure 4.29).

Figure 4.29: Mergin Maps Plugin Project Context Menu

15. When prompted open the project file. You will see the data you just finished collecting.

5. Advanced Project Configuration

This chapter will cover more specific and customized settings used by many data collection projects. These are techniques you can use to avoid data entry errors and make your data collection forms more intuitive and intelligent.

This chapter includes the following tasks:

- **Task 1 - Advanced Form Configuration**
    - Default Values
    - Constraints
    - QR Codes
    - Configuring forms with text, HTML and spacer widgets
- **Task 2 - Advanced Photo Settings**
    - Configuring a custom photo folder
    - Configuring custom photo file names
    - Accessing EXIF data
- **Task 3 - Working with Non-spatial tables**
    - Drill down / Cascading forms
    - Collecting multiple photos per feature
    - 1-N Relations

After this chapter you will be able to:

- Set constraints on fields
- Hide / show fields based on values
- Use form design widgets (text/HTML/spacer)
- Configure a custom photo folder
- Configure custom photo file names

- Access EXIF data
- Configure drill down / cascading forms
- Attach multiple photos per feature
- Work with 1-N Relations

5.1   Task 1 - Advanced Form Configuration

There are additional form settings which can improve workflows and allow you to create more intelligent data collection forms. For example, in Chapter 2 you learned to use default values for fields. This included the use of variables and expressions. This section will focus on several useful settings including using Constraints, Hyperlinks, setting Field visibility and using form configuration widgets such as Text, HTML and Spacers.

Default Values

The Default values you configured in the Configuring Survey Layers chapter were for a horizontal accuracy field and a date field. Here you will learn one more use case, transforming X and Y coordinates into a specific coordinate system. You will use a project designed to conduct bird surveys named `qmm-bird-survey`.

1. First, clone the project using the Mergin Maps dashboard (app.merginmaps.com as described in the previous chapter in Task 6 - Collecting Data with Multiple Survey Layers, on page 58.

2. Open QGIS and download the cloned project from the Mergin Maps Data provider in the *Browser panel*.

3. The project has one survey layer and a background layer. Open *Layer Properties* for the birds layer. Switch to the *Attributes Form* tab.

4. There are two fields for coordinates of the observation X and Y.

5. The birds layer and the QGIS project are in the EPSG:3857 CRS. This is WGS 84 Pseudo-Mercator and it is used by most streaming basemaps such as Google Maps, OpenStreetMap, Bing etc. Here you will configure this field to register the coordinates in latitude and longitude (EPSG:4326).

6. Select the X field to view the widget settings. In the *Defaults* section, find *Default value*. You will enter an expression. Click on the *Expression Builder* button to open the Expression Builder.

7. You will use the X() function which just requires the geometry of each feature. You can use the variable $geometry for the geometry value (figure 5.1). You will then use the transform() function to transform the coordinate from EPSG:3857 to EPSG:4326. The final expression should look like:
x(transform($geometry, 'EPSG:3857', 'EPSG:4326')).

Figure 5.1: X Default Value Expression

8. Check the *Apply default value on update* option. This will ensure the default value is applied anytime the feature is edited.

9. Repeat these steps for the Y field using the Y() function.

10. When you survey a new point in Mergin Maps mobile app, you will see the values are automatically filled in as latitude and longitude (figure 5.2, on the following page).

Constraints

Another method of reducing data collection errors is using Constraints. These allow you to limit what can be entered as values into a field during data collection. For example, you can ensure that a value is entered by enforcing *not null* constraints. In that situation, if no value is entered, the data collector will get a warning. You can also use an expression to specify a range of valid values. Here you will go through one example of using constraints using the same bird survey project used in the previous example. One of the items to be collected is the number of birds present at an observation. Since it makes no sense to

Figure 5.2: Latitude and Longitude Values in the Mergin Maps Mobile App

record a point, if the number is not at least one, you will set a constraint on the number field.

1. Open *Layer Properties* for the birds layer.

2. In the *Attributes Form* tab, select the number field in the *Available Widgets* column on the left.

3. In the *Widget Display*, fine the *Constraints* section.

4. For the *Expression* type "number" >= 1 and check the *Enforce expression constraint* option. This makes it a *hard constraint*, meaning the data collector cannot enter zero for this field.

5. Check the *Not null* and *Enforce not null* constraints options. By itself the *Not null* constraint is considered a *soft constraint*. When unmet, the data collector will get a warning but can proceed. Checking the *Enforce not null* constraint option, makes it a *hard constraint* and requires a value be entered before the feature can be saved.

6. Your widget settings for the number field should look like figure 5.3, on the next page.

7. When surveying new features in Mergin Maps mobile app, this field will

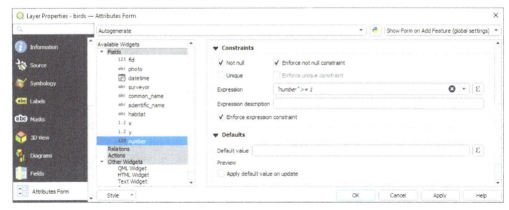

Figure 5.3: Constraint Settings for the Number Field

have to be filled in using a value higher than or equal to 1. Otherwise, you will get a warning (figure 5.4).

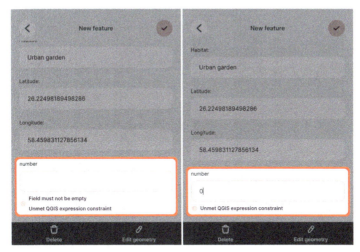

Figure 5.4: Not Null (left) and Number (right) Constraints in the Mergin Maps Mobile App

8. Other common uses of Constraints include using regular expressions to limit the type of characters which can be entered into a field. For example, this expression, `not regexp_match(common_name,'[^A-Za-z]')`, ensures that the field `common_name` can only accept alphabetical, and not numeric characters.

QR Codes

It is possible to scan QR codes with Mergin Maps and have the associated text (e.g. a link) be filled in automatically.

It is quite simple. All that is required is for the field name or Alias to have the word QRCode in it. It can be upper, lower or mixed case.

When used in the Mergin Maps mobile app, there will be a QR code icon next to the field on the form (figure 5.5). Click on the QR code icon to scan the QR code using your camera. The associated text will be filled in as the field value.

Figure 5.5: QR Code Icon Next to Field in the Mergin Maps Mobile App

This project is available as a public project named *qmm-monuments-qrcodes*.

Configuring forms with text, HTML and spacer widgets

There is a suite of widgets which do not necessarily pertain directly to attribute fields, but can be used to improve the design of, and add information to, a form. These are the **Text**, **HTML** and **Spacer** widgets. These can be used with any data collection form. Therefore, this section will simply describe their configuration and provide some use cases.

These are all found in *Layer Properties* on the *Attributes form* tab under *Available Widgets* in the *Other Widgets* section when using the *Drag-and-Drop Designer* (figure 5.6).

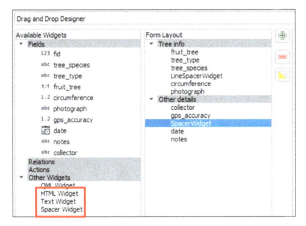

Figure 5.6: Text HTML and Spacer Widgets in the Drag and Drop Designer

Spacer Widgets

The Spacer widget can be useful if you want to have some space between the fields in your form or add a horizontal line to help break the form up into sections. The default setting simply provides extra space between items on a form. There is also an option to *Draw a horizontal line* (figure 5.7). To add one, to a form simply drag and drop it where it should go on your form. At that time the configuration options will open.

Figure 5.7: Spacer Widget Options

Figure 5.8: Spacer Widget on Mergin Maps Mobile App

Text Widgets

Text widgets can be used to add instructions based on a simple text string a form. They can also be used to add additional information, since they can be configured with expressions. For example, information can be presented based on data entered into the form.

Here is an example of using entered values along with some instructions. The form is for conducting a vehicle inventory. One item being collected is the vehicle registration number (license plate). It is important that this be visible in the photograph taken. To ensure this is done each time, a Text widget will be configured as a reminder. It will be set up so the entered value for the Vehicle Registration Number field is included in the note.

1. To configure a *Text widget,* drag it onto the form using the *Drag-and-Drop Designer*. The properties for the widget will open.

2. You can begin by typing any text string into the box. For example, Make sure the number plate is visible in the photo. This is how you can configure a

widget with simple text.

3. Here we will add the plate number entered in the "VRP" field. We will put our cursor after the word *plate* and choose the VRP field from the dropdown list, then click the green  button to add it. Fields are added in this format [% "VRP" %].

4. Now the widget is configured with the following expression: Make sure the number plate [% "VRP" %] is visible in the photo. (figure 5.9).

5. Click *OK* to accept the settings.

Figure 5.9: Text Widget Configuration in QGIS

6. After saving and synchronizing the project, this is how it looks during the field survey (figure 5.10).

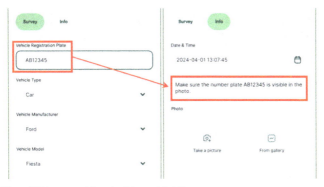

Figure 5.10: Text Widget on Mergin Maps Mobile

HTML Widgets

HTML widgets embed an HTML page into a form. These can be used to display online images and other online resources such as PDF files, videos and websites.

To configure an HTML widget:

1. Using the Drag-and-drop designer, add a HTML Widget to your form.

2. Double-click on the widget to configure it.

3. Click the expression button and enter your expression. The following expression is a good template for displaying an online image in a form:

```
'<a href=\\"'||attribute( @feature, 'link' )||'\\"><img src=\\"'||
attribute( @feature, 'link' )||'\\" width=300></a>'
```

In this example, your form should contain a text field for storing the full URL link. Here the field is named link.

4. Click *OK* to accept. Back at the Widget configuration window click the green ⊕ button to add the expression. QGIS automatically applies HTML formatting and functions to evaluate your expression, resulting in following code:

```
<script>document.write(expression.evaluate("'<a href=\\"'||attribute(
@feature, 'link' )||'\\"><img src=\\"'||attribute( @feature, 'link' )
||'\\" width=300></a>'"));</script>`
```

5. Click *OK* to accept the settings.

6. The following code '<a href="'||attribute(@feature, 'link')||'">here is your link</a>' can be used to embed a link in a form. Again, your form should contain a text field for storing the full URL link. In this example, the field is named link.

5.2   Task 2 - Advanced Photo Settings

In Chapter 3 you learned the basics on how to set up a field data collection form in QGIS to be able to take photos. There are several additional important settings for photos such as creating a specific folder for photos, controlling image size and customizing photo file names. It is also important for many workflows to be able to take multiple photos per feature or to be able to access image EXIF data. These topics will be covered here.

Configuring a custom photo folder

It can be useful for a larger project to organize photo files into a specific folder. Otherwise they are simply dumped into the project folder. It can also be desirable if you are using Selective Synchronization because one option is to apply Selective sync to a folder.

1. Create a subfolder in the Mergin Maps project folder. In this example it is named photos.

2. Open your QGIS project.

3. This example is using the bird survey project used at the beginning of the chapter and the photos attachment widget has already been configured (figure 5.11, on the following page).

4. Next you will change the *Default path* to the photos folder. Click on the *Data defined* override button ⬅️ and choose *Edit* from the context menu (figure 5.12, on the next page).

5. In the *Expression Builder* use the search box to find the variable @project_folder and double-click on it to add it to the expression. Then add a plus sign and the string '/photos' (figure 5.13, on page 77). Click *OK* when finished.

6. The data defined override for the *Default path* is now set to your expression, signified by the yellow expression data defined override icon ε (figure 5.14, on page 77).

7. Click OK to apply the changes and save your project. Sync the project and your photos will now be saved into the photos folder.

Configuring custom photo file names

Using naming conventions for GIS files has always been a good practice. The same applies to photographs collected in the field. Using the Mergin Maps QGIS plugin you can control the photo file names using expressions.

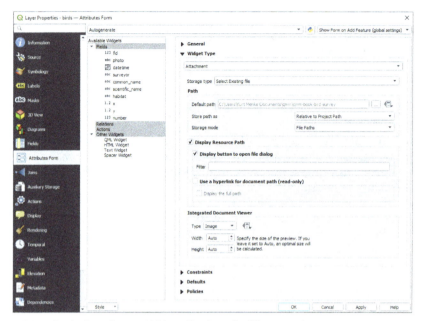

Figure 5.11: Basic Attachment Widget Settings for Photographs

Figure 5.12: File Path Data Defined Override

1. Open your QGIS project and open from the *Project* menu open *Project Properties*.

2. Click on the *Mergin Maps* tab and find the *Photo attachment configuration* section.

3. Your survey layer(s) will be listed. Select one, and click the *Expression Builder* button.

4. In the *Expression Builder* you can write an expression which will be used as

Figure 5.13: Custom Photo Folder Data Defined Override

Figure 5.14: Active Data Defined Override for Photo Path

the photo file name.

5. The variables @layer_name and @mergin_username are often used in combination with a datestamp format_date(now)).

6. For example, @layer_name || '-' || @mergin_username || '-' || format_date(now(),'yyyy-MM-dd-hh-mm-ss').

7. Click *OK* to apply. If you have multiple survey layers, you can set up individual custom naming expressions for each.

8. A preview result is displayed in the Project Properties --> Mergin Maps (figure 5.15, on the following page).

9. There are some best practices to keep in mind when creating expressions for custom photo file names.

    a. Each photo needs to have a unique name to avoid synchronization issues. When more than one surveyor is deployed, use of the variable @mergin_username is recommended. Even if

Figure 5.15: Custom Photo File Name Expression

photos are taken at the same timestamp, this will differentiate them. This will also make it easier to sort photos later based on who took them. Likewise, if you have multiple survey layers using the variable @layer_name is recommended.

b. The file extension (jpg, png etc.) is added automatically.

c. If you use a field value in the expression, make sure that this field will be filled out during the survey. You can use **Constraints** to manage this.

d. If you want to use a value from a numeric field in the photo name, you must first convert it to a string using the to_string() function.

e. The photo file is named via the expression, with the **current** field values, when the photo is taken. Therefore, the photo names will not change if the values are changed during a later editing session.

f. Like anything done with Mergin Maps, the configuration must be saved and synchronized. Only photos taken afterwards will be named via the expression.

10. Here are some example expressions that can be used or modified to fit your needs:

- Expression: @layer_name + '-' + @mergin_username + '-' + format_date(now(),'yyMMddhhmmss').
  - Preview: birds-sarah-240707154052.jpg
  - Description: This is a combination of the name of a layer (birds), Mergin Maps username (sarah)

and reformatted timestamp that starts with the year and ends with seconds.

- Expression: `"species" + format_date( now(),'-yyyyMMdd-hhmmss')`.

  - Preview: `Blackbird-20240707-154052.jpg`
  - Description: Blackbird is a value of the species field. Current timestamp is reformatted with added hyphens to separate the date and time.

- Expression: `'photo-' + format_date( now(),'ssmmhhddMMyy')`.

  - Preview: `photo-520415070724.png`
  - Description: A string can be added to the photo name (here: photo-). The order of the timestamp is reversed (compared to the previous examples), starting from seconds.

- Expression: `@layer_name + ' ' + to_string("house-number") + ' at ' + format_date( now(),'ssmmhh') + ' on '+ format_date( now(),'ddMMyy')`.

  - Preview: `house 41 at 520415 on 070724.png`
  - Description: Here we use the name of a layer (house), followed by a string adding space. A numeric field (house-number) is converted to a string. The timestamp is divided to display the time and date separately, with added strings at and on to make the photo name more readable.

Accessing EXIF data

EXIF data is essentially metadata for photographs. It is an acronym which stands for Exchangeable image file format. Every time you take a photo with a digital camera or a smartphone, information is saved along with the image itself. It can include details of the camera, lens and shooting settings used, plus optional information about the photographer, location and more.

The Mergin Maps mobile app supports some expression functions that can be used to read EXIF metadata and store their values in the fields of your survey layer.

> To store GPS (location) EXIF metadata, both the mobile app and your camera app must have location permissions enabled on your mobile device.

Below are the steps to store EXIF metadata values in the fields of your survey layer.

1. In QGIS, open the *Layer Properties* of your *survey* layer and navigate to the *Attributes Form* tab.

2. You will simply use an expression to define the *Default values* of fields that should store EXIF metadata.

3. Supported EXIF functions are listed below. In general, EXIF functions look like this: `read_exif('<ABSOLUTE_PATH_TO_IMAGE>', '<EXIF_TAG_STRING>')`.

4. For example, the default value expression for the direction of the image (EXIF tag GPSImgDirection) can be defined as follows: `read_exif(@project_home + '/' + "photo", 'GPSImgDirection')`

5. This expression requires the absolute path to an image. The absolute path is defined using the field where the image is stored (here: `photo`) and the `@project_home` variable that refers to the project home folder.

6. The preview shows a warning *Function is not known*. This is OK, it is not a QGIS function, but the **Mergin Maps mobile app** will know what to do with it (figure 5.16, on the next page)!

7. Here is a list of EXIF functions supported by the Mergin Maps mobile app:

- `read_exif_img_direction('<ABSOLUTE_PATH_TO_IMAGE>')` returns the direction of the image when captured. The direction is represented by a number in degrees from 0 to 360.
- `read_exif_latitude('<ABSOLUTE_PATH_TO_IMAGE>')` returns GPS Latitude as a decimal number.
- `read_exif_longitude('<ABSOLUTE_PATH_TO_IMAGE>')` returns GPS Longitude as a decimal number.
- `read_exif('<ABSOLUTE_PATH_TO_IMAGE>', '<EXIF_TAG_STRING>')`, where EXIF tag string defines the EXIF property, such as:

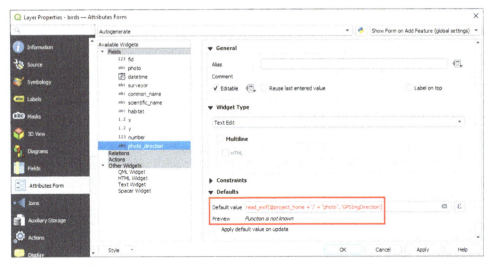

Figure 5.16: EXIF Direction as a Default Value for a Survey Field

- GPSImgDirection returns the direction of the image when captured as a rational. The direction is represented by a rational, e.g. 350/1.
- GPSLatitude returns GPS Latitude as rationals, e.g. 48/1, 6/1, 309275/10000.
- GPSLongitude returns GPS Longitude as rationals, e.g. 17/1, 6/1, 244907/10000.
- ImageWidth returns the image width in pixels.

The list of EXIF tags can be found in the Android developer documentation for *ExifInterface*[a] or in *ExifTool* documentation.[b] Note that some of the tag names listed in *ExifTool* documentation can differ from the EXIF specification. As an example, while *ImageWidth* tag is valid, *ImageHeight* is not, and you have to use *ImageLength* tag name (defined by EXIF specification) when recording EXIF metadata in the *Mergin Maps mobile app*.

[a]https://developer.android.com/reference/android/media/ExifInterface
[b]https://exiftool.org/TagNames/EXIF.html

5.3 Task 3 - Working with Non-spatial tables

As you have learned, Mergin Maps supports multiple vector survey layers per project and those survey layers can have any vector geometry: point, line and polygon. Mergin Maps also supports non-spatial tables and these can be an important part of a survey project. Non-spatial tables can be used either on their own or they can link to a spatial layer. In this section you will learn how to use them in Value Relation widgets (drill down forms). You will also learn how to implement them if you need to take multiple photos per feature or if you need to revisit the same sites on a regular basis. In the latter scenario, it is better to add new data with each visit rather than recollecting the same points each time. These situations involve 1-N relations.

Drill down / Cascading forms

These allow you to create a drop-down list related to a value entered into another field. To demonstrate this, you will work with a vehicle survey. The data collector will be asked to identify the vehicle type from a drop-down list. Based on the selection (car, truck, bus, motorbike), a filtered set of vehicle manufacturers and models will be available for subsequent questions. There is a single survey layer *vehicle* along with 5 non-spatial tables. The cascading forms will be built using **Value Relation** widgets for several fields. By configuring these widgets against the non-spatial tables, the values in the drop-down menus will be filtered based on the previously selected values. For example, if the Vehicle type Cars is selected, the Vehicle-Manufacturer field will offer only car manufacturers: Audi, Ford, Mercedes-Benz, Skoda, Tesla and Toyota.

1. Using the steps outlined at the beginning of the chapter, find and download the *qmm-drilldown-forms* project using the Mergin Maps data provider in the *Browser panel*. It's a public project.

2. Open each non-spatial lookup table to familiarize yourself with their contents. They are simple tables.

3. The *vehicletype* table contains two fields, an `fid` and a `name` field with names of different types of vehicles.

4. Open *Layer Properties* for the *vehicles* layer and switch to the *Attributes Form* tab. Notice that the form has already been set up in the *Drag and Drop Designer* and some basic settings. You will configure the value relation for the **Vehi-**

cle Type, **Vehicle Manufacturer**, **Vehicle Model**, **Colorized** fields, which are located in the *Survey* tab of the form.

5. Select the vehicle_type field. This will relate to the *vehicletype* table.

6. Switch the *Widget type* to *Value Relation* and configure it as follows:

   a. Set the *Layer* to *vehicletype*. *vehicletype* refers to the non-spatial table which has the fid and name fields.

   b. Set the *Key Column* to fid. This establishes the relationship between this field and the *vehicletype* table.

   c. Set the *Value Column* to name. This column will provide the dropdown values for the vehicle_type field.

   d. Uncheck *Allow NULL value*. This will limit the dropdown to the data in the *vehicletype* table (figure 5.17). Checking *Allow NULL value* will add a *(no selection)* option to the dropdown list.

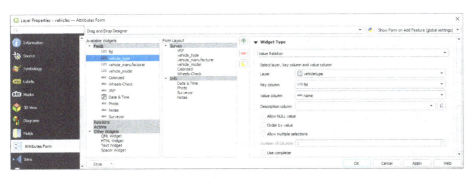

Figure 5.17: The Vehicle Type Value Relation

7. At this point if you close *Layer Properties*, put the layer into edit mode and try adding a point, you will see the first Value Relation widget in action. The values from the *vehicletype* non-spatial table are populating the pick list (figure 5.18, on the next page).

8. Open *Layer Properties* and select the vehicle_manufacturer field. This will relate to the *vehiclemanufacturer* table, which has the fields fid, name, and type. Switch the *Widget type* to *Value Relation* and configure it as follows:

   a. Set the *Layer* to *vehiclemanufacturer*.

Figure 5.18: The Vehicle Type Value Relation in Form View

b. Set the *Key Column* to fid. This establishes the relationship between this field and the *vehicletype* table.

c. Set the *Value Column* to name. This column will provide the dropdown values for the vehicle_type field.

d. Set the *Filter expression* to "type" = current_value('vehicle_type'). Here you are using the current_value() function which returns the current value entered in the form. This will limit the dropdown list to just those for the entered vehicle type.

e. Uncheck *Allow NULL value*. This will limit the dropdown to the data in the *vehiclemanufacturer* table (figure 5.19).

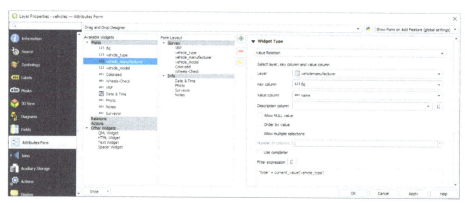

Figure 5.19: The Vehicle Manufacturer Value Relation

9. Now, the form will offer only the options, where the vehicle_type field of the *vehiclemanufacturer* table matches the current value of the Vehicle Type

field. When entering data, you will get a drill-down form. For example, when *Truck* is selected as the *Vehicle Type*, only the six truck manufacturers are available in the Vehicle-Manufacturer field (figure 5.20).

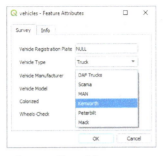

Figure 5.20: The Vehicle Manufacturer Form View

10. Next is the *vehicle-model* field which will work similarly. It will be configured to refer to the *vehiclemodel* table, limiting options based on the vehicle-manufacturer field.

11. Switch the *Widget type* to *Value Relation* and configure it as follows:

   a. Set the *Layer* to *vehiclemodel*.

   b. Set the *Key Column* to fid.

   c. Set the *Value Column* to name.

   d. Use the manufacturer field to filter the values using this *Filter expression*:
   "manufacturer" = current_value('vehicle_manufacturer').

   e. Uncheck *Allow NULL value* (figure 5.21, on the following page).

12. Next you can configure the *Colorized* field. Follow the same steps as you did for the vehicle_type field. Set the *Key Column* to fid and the *Value Column* to name. No *Filter expression* is needed because this won't be filtered by other selections. But you can check the option *Allow multiple selections*.

13. Finally, you can configure the *Value Relation* widget for the Wheels-Check field.

   a. Set the *Layer* to *wheel-check*.

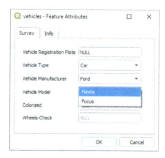

Figure 5.21: The Vehicle Model Form View

b. Set the *Key Column* to fid.

c. Set the *Value Column* to wheel.

d. Check *Allow NULL value* and *Allow multiple selections*.

e. For the *Filter expression* enter:
   "type"= current_value('vehicle_type') (figure 5.22).

Figure 5.22: The Final Vehicle Form View

Collecting Multiple Photos per Feature

There are many instances where taking more than one picture per feature is necessary. This can be accomplished via a 1-N relation.

The workflow involves creating a non-spatial table which has a field for the path to the photos, and a field for the survey layer primary key (external-pk). You then add a unique-id field to your survey layer with a default value of uuid(). Next, create a relation via *Project Properties* between the non-spatial table (external-pk) and the survey layer (unique-id).

Finally, in the *Attributes form* tab in the *Layer Properties* of the non-spatial table, set the *Widget Type* of the external-pk field to *Relation Reference*. Mergin Maps mobile app detects the type of 1-N relation and displays the image viewer for the relations.

The following task will go through this procedure step-by-step.

To link multiple photos to a single feature, you need a unique field to link following tables:

- Survey layer containing spatial information
- A non-spatial table containing path to the photos

1. This example will use the bird survey project. It contains a single survey layer, a non-spatial table and a background map.

2. To set 1:N relation between these tables correctly you need to add a new text field to the survey layer. Open *Layer Properties* for the birds survey layer and switch to the *Fields* tab. Click the *Toggle edit mode* button to put the layer into

edit mode. Click the *Add new field* button. Name the field unique-id and make it of *Type Text (String)* (figure 5.23).

Figure 5.23: Adding the Unique-id Field

3. Toggle out of edit mode and save your edits.

4. Switch to the *Attributes Form* tab and select the new unique-id field. Use the function uuid() as the *Default value*. This function assigns a unique identifier to every created feature, even when different surveyors create features simultaneously (figure 5.24).

Figure 5.24: UUID() as the Default Value

5. Now you need a field in the photos table that will be used to store the UUID of features from the survey layer (the foreign key). Open *Layer Properties* for the photos table and switch to the *Fields* tab. Repeat the steps above to add a new text field in this table named external-pk. Toggle out of edit mode and save your edits. Close *Layer Properties*.

6. Now you will configure the 1-N relation. From the *Project* menu choose *Properties* to open *Project Properties*. Switch to the *Relations* tab.

7. Click the *Add Relation* button. You will define the parent and child layer and the fields to link these layers (figure 5.25, on the facing page):

   a. Name is the name of the relation and it can be left blank.

   b. *Referenced (parent)* is the spatial layer birds.

c. *Field 1* of the *Referenced (parent)* is the field `unique-id` that contains the UUID.

d. *Referencing (child)* is the non-spatial table `photos`.

e. *Field 1* of the *Referencing (child)* layer is the `external-pk` that contains the foreign key to link photos with surveyed features.

f. Click *OK* when finished.

Figure 5.25: Adding the 1-N Relation

8. Open *Layer Properties* for the photos table and switch to the *Attribute Form* tab. Select the `external-pk` field. Set the *Widget Type* to *Relation Reference* (figure 5.26).

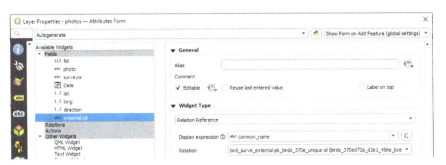

Figure 5.26: Relation Reference Widget Setup

9. Apply the changes and close *Layer Properties*. Save the project and synchronize.

> This project is available as a public project named *qmm-bird-survey-multiple-photos-solution*.

1-N Relations

It is often the case that you have a set of spatial features and you need to revisit them on a regular basis to collect data about their status. For example, there is a GIS layer representing the trees in public spaces within a municipality. Surveyors visit each tree twice per year to conduct a risk assessment where their structural integrity is assessed. Instead of duplicating the tree layer or collecting the same position over and over, inspections can be recorded in a non-spatial table that is linked to the spatial layer. This way, multiple records can be linked to one feature.

> This workflow can be paired with *Staking out* points, to easily navigate to each. This is covered in Task 3 of the Field Survey chapter.

1. This public survey project (qmm-book-risktree-survey) contains one point layer, a non-spatial table and several basemaps.

2. The survey layer has a simple array of attributes (figure 5.27). uuid values are generated using uuid() function as a default value when a feature is created. This ensures that these values are unique even when multiple surveyors capture new features at the same time. This field will be used to link risk surveys and trees.

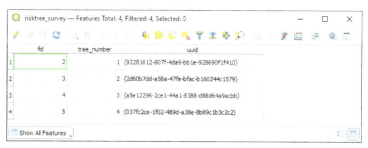

Figure 5.27: Survey Layer Attributes

3. The real information is stored in a non-spatial table (figure 5.28, on the next page). This includes all the important information from the risk tree survey.

`Tree_id` is filled in automatically based on a 1-N relation that you can set up in QGIS.

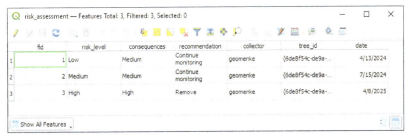

Figure 5.28: Non-spatial Table Attributes

4. To configure the 1-N relation in QGIS (figure 5.29, on the following page):

   a. From the main menu, select *Projects | Properties*.

   b. In the *Relations* tab, select *Add Relation*. Name is the name of the relation and it can be left blank.

   c. *Referenced (parent)* is the spatial layer `risktree_survey`.

   d. *Field 1* of the *Referenced (parent)* is the field `uuid` that contains the UUID.

   e. *Referencing (child)* is the non-spatial table `risk_assessment`.

   f. *Field 1* of the *Referencing (child)* layer is the `tree-id` which acts as a foreign key to link surveys to the spatial features.

   g. Click *OK* when finished.

5. The last configuration step involves the `risktree_survey` Attribute form. Open *Layer Properties* for the `risktree_survey` layer, and select the *Attributes form* tab.

6. Drag and drop the *Risk Survey Relation* to the *Form Layout*. Because the relation has been established in the project, the *Widget* automatically changes to a *Relation Reference* (figure 5.30, on the next page). Click *OK* to accept.

7. Now the survey team can add multiple inspections for each risk tree. The survey records will be stored in the `risk_assessment` table. When you open the form for an existing record in the `risktree_survey` point layer, it will display

Figure 5.29: Configuring the N-1 Relation

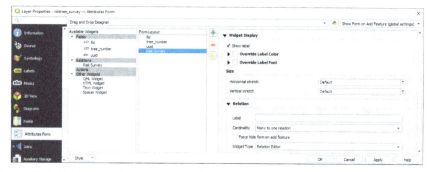

Figure 5.30: Relation Reference Widget Setup

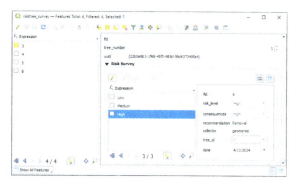

Figure 5.31: Form View of Survey Data for a Single Tree

existing inspection records and you can also add, delete or edit the records
(figure 5.31).

The Mergin Maps project for this task is public and you can download it if you want to check your settings and inspect the solution. It is named *qmm-book-risk-tree-survey*.

6. Project Management and Collaboration

In this chapter we will look more into the project management and collaboration with other users.

One of the main features of the Mergin Maps platform is the synchronization between different devices, as we have already seen in the previous chapters. However, this also means that multiple users can work with the same project at the same time and synchronize their changes back to the cloud. Mergin Maps tries the best to automatically and safely merge these edits.

This chapter includes the following tasks:

- Task 1 - Create a public Mergin Maps project
- Task 2 - Add users to your workspace or project
- Task 3 - Resolve a conflict between versions
- Task 4 - Deploy a new project version

After this exercise you will be able to:

- Make your project public/private.
- Collaborate with other users and manage their access to the project.
- Understand Mergin Maps member roles and permissions.
- Avoid synchronization issues.

6.1 Task 1 - Create a public Mergin Maps project

You have worked with multiple public projects in the previous chapters. In this task you will create a public project and explore options and limitations of sharing it with others.

1. Open a new (empty) project in QGIS. Click on the *Create Mergin Maps Project* button ⊞ in the Mergin Maps toolbar.

2. Select the *New basic QGIS project* option.

3. Enter the name of your project and the path to the folder where it will be saved. Check the *Make the project public* option (figure 6.1). Click *Finish* to create the project in your workspace.

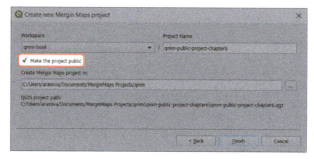

Figure 6.1: Creating a Public Mergin Maps Project in QGIS

4. Go to the Mergin Maps dashboard (app.merginmaps.com) and sign in.

5. In the *Projects* tab, find your new project and click on it. Go to *Settings*. Note there is a *Make private* button that can be used to make the project private. (figure 6.2).

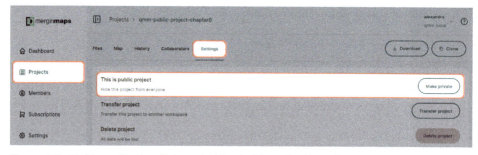

Figure 6.2: Making a Project Private in Mergin Maps Dashboard

Other Mergin Maps users can now find your public project when exploring public projects through Mergin Maps dashboard, mobile app or QGIS. Another option is to share a URL link to the project. The URL can be copied from your browser (figure 6.3). The project can be downloaded by anyone with this link, even if they do not have a Mergin Maps account.

Figure 6.3: Mergin Maps Project URL

Public projects are accessible as read-only. This means others can download the project and data, but cannot actively contribute by surveying new features or making changes in the project settings (to do this, they need to have appropriate roles or permissions as explained in the next task). Making a project public is useful to share a project with people who are not registered Mergin Maps users or to share interesting projects or project templates with the community.

> Mergin Maps provides several public projects that demonstrate how to use various form widgets or Mergin Maps features. When browsing public projects, search for the *documentation* keyword.

6.2 Task 2 - Add users to your workspace or project

In this task, we will add another user to the project. When collaborating with other people, it is important to define their member roles and permissions.

On the workspace level, users can be added as *members* or as *guests*. Members automatically gain access to all projects in the workspace. Their *member role* defines the default level of access to the projects in the workspace and to the workspace itself. This can save some time as there is no need to manually manage their access for every single project. *Guests* gain access only to specific projects they are invited to.

Member roles and *project permissions* use similar terms to define what users can or cannot do. Member roles define these actions on the workspace level. Project permissions are given on a project level.

These roles and permissions are as follows:

- *Reader* can see projects in the workspace, project data and history.

- *Editor* can add, edit and delete features, but cannot make changes in the project settings (symbology, forms, project properties, etc.).

- *Writer* can also add, edit and delete features as well as change project settings.

- *Admin* can also create new projects, manage workspace members, delete projects or transfer them to another workspace. This role is only available for workspace members.

- *Owner* has also access to the invoicing and subscription. Owners has full access to the project or to the workspace.

Project permissions can be defined in addition to the roles of workspace members: for instance, a workspace member with the reader role can be a writer or an admin for a specific project.

Member roles and permissions are important especially when working in larger or complex teams. *Readers* cannot do any changes, so this role is appropriate for those who need to see the data (e.g. the results of your survey), but will not actively contribute. *Editor* is the most appropriate role/permissions for field surveyors as they can add and edit data either in the field or in QGIS. *Writers* can also modify the data and project settings, so these users should have some basic skills in terms of QGIS and the Mergin Maps mobile app. The roles of *Admin* and *Owner* should be reserved for those who manage the user access, workspaces, projects and subscription. Ideally, it should be one person or a small number of highly coordinated people.

In this task, we will invite a user named *Sarah* to become a guest with an access to two projects. She will get the *writer* permission, meaning she can add, edit and delete features as well as make changes in the project settings.

1. Log in to the Mergin Maps dashboard (app.merginmaps.com).

2. Navigate to the *Members* tab in the left panel. Click on the *Invite* button (figure 6.4, on the facing page).

3. Fill in the invite form (figure 6.5, on the next page):

- Enter an e-mail of the person you want to invite. Multiple e-mails (people) can be invited at once.

Figure 6.4: Members Tab in Mergin Maps Dashboard

- Choose the *Guest Workspace role*.

- From the list, select projects that you want to share. Multiple projects can selected.

- Choose the *writer Project permission* from the list.

Figure 6.5: Inviting New Collaborators to Mergin Maps Workspace

4. Use the *Invite* button to send the invitation.

The invited person will get an e-mail with a link. After they confirm the invitation, they will be a *guest* in the workspace and have *write* permission to the specified project.

The overview of members and guests in a workspace can be found in the *Members* tab of the dashboard (figure 6.6). It is accessible only to owners and admins of the workspace who can grant, change and revoke user roles and permissions.

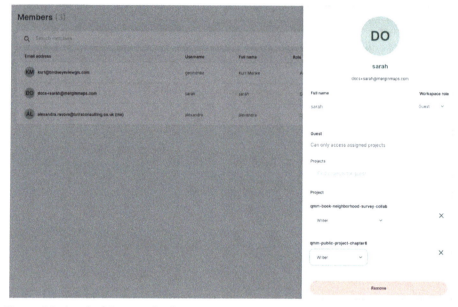

Figure 6.6: Level of Access of a Guest in a Mergin Maps Workspace

> Private projects can also be shared using a URL in the same way as public projects (see Task 1). The recipient can request access to the project after signing in. Owner of the workspace will be notified and can grant the access in the Mergin Maps dashboard.

6.3   Switching between workspaces

A Mergin Maps user can have access to multiple workspaces. Projects are tied to a specific workspace, so the appropriate workspace needs to be selected first in order to work with a particular project. Here we will show how to switch between workspaces in different Mergin Maps components.

Switching to another workspace in Mergin Maps dashboard

The current workspace is displayed in the upper right corner of Mergin Maps dashboard (app.merginmaps.com) under the user name.

To switch to another workspace, click on the user name. Recently used workspaces are listed under the *Workspaces* section (figure 6.6, on the preceding page). Click on the workspace you want to switch to and the dashboard will reload the content of the workspace.

Figure 6.7: Switching Workspaces in Mergin Maps Dashboard

Alternatively, click on the *Manage workspaces* option to display all workspaces available to you and the overview of their details. From here, you can also leave a workspace if you no longer wish to participate.

Switching to another workspace in QGIS

In QGIS, navigate to the *Browser panel* and right-click on the *Mergin Maps* provider. The *Switch workspace* option opens a list of available workspaces (figure 6.8, on the following page). Choose one and click on the *Select workspace*. The content of the *Mergin Maps* provider will be reloaded and the projects from the selected workspace will be displayed. These projects can be then downloaded and opened in QGIS.

Switching to another workspace in Mergin Maps mobile app

Tap the account icon  in the upper right corner of the main page in Mergin Maps mobile app. The current workspace is displayed under *Workspaces*

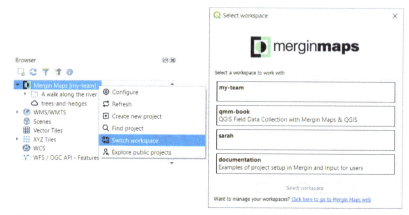

Figure 6.8: Switching Workspaces in QGIS

section. Tap on it to open a list of available workspaces (figure 6.9). Select a workspace from the list. Projects from this workspace will be displayed in the Projects [Projects] tab on the main page of the mobile app and can be downloaded and opened in the app.

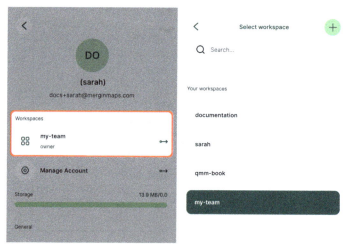

Figure 6.9: Switching Workspaces in Mergin Maps Mobile App

6.4   Task 3 - Resolve a conflict between versions

In the previous chapters, we have already synchronized projects: a project created in QGIS was synchronized to Mergin Maps cloud, downloaded to the mobile app and changes made in the mobile app were synchronized back to the cloud. Mergin Maps takes care of dealing with the changes by comparing the original and the modified datasets and merging them together.

Synchronizing changes made by different users or devices

Here we will use a slightly modified project from previous chapter to demonstrate how Mergin Maps synchronization works. It is a public project named qmm-book-neighborhood-survey-collab and can be downloaded using this link.[6]

Fields updated_by and updated_at were added to all layers to record who and when did the last update. These fields use default values @mergin_username and now() with the *Apply default value on update option* so they are filled out automatically when someone modifies the feature (as described in Default Values, on page 66).

There are two surveyors working simultaneously: alexandra is surveying parks using the Mergin Maps mobile app, while sarah is mapping parks in QGIS. After finishing their work, they synchronize their changes to Mergin Maps: alexandra is the first one to sync the changes, so when sarah synchronizes her changes in QGIS, she sees that there are some *Server Changes* in the *Project status* window (figure 6.10, on the next page).

The synchronized dataset contains features captured by both surveyors (figure 6.11, on the following page). Note that the fid of features surveyed by sarah were changed, as fid 2 and fid 3 were already assigned to features in the parks layer surveyed by alexandra that were pushed to the cloud first. Mergin Maps automatically assigned the next numbers in the sequence to fid for features synchronized subsequently.

To become familiar with the synchronization process, recreate this process in your project:

1. Open your Mergin Maps project in QGIS. Add some new features to your

---

[6]https://app.merginmaps.com/projects/qmm-book/qmm-book-neighborhood-survey-collab/

Figure 6.10: Project Status Showing Both Server and Local Changes in the Project

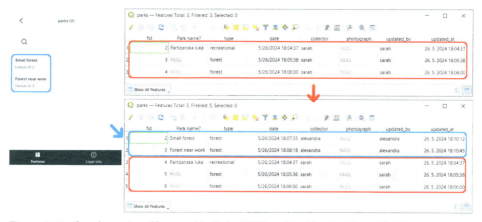

Figure 6.11: Synchronizing Changes Made in QGIS and in Mergin Maps Mobile App. Synchronized Dataset Contains Features from Both Sources.

survey layers or change something in the project setup (e.g. symbology of a layer).

2. Open the same project in Mergin Maps mobile app. Add some new features. Synchronize the changes using the [Sync] button.

3. Go back to QGIS and synchronize the changes using the button in Mer-

gin Maps toolbar. Notice the *Server Changes* and *Local Changes* in the *Project Status* window. Features collected using the mobile app should appear in the map window.

4. Go back to the mobile app and synchronize again to download the synchronized dataset and observe the changes.

Synchronization conflicts and resolving conflict files

Sometimes changes cannot be put together automatically. In these cases, Mergin Maps creates a conflict file that can be inspected and can be used to retrieve the correct version of the project. Conflicts can occur when there are changes in the data schema, such as deleted fields or changed data types. Multiple surveyors modifying the same feature differently at once can also produce conflicts. In this case, Mergin Maps keeps the latest version, but keeps track about the other versions in a conflict file.

Here, we will create a conflict file intentionally. This is something **you want to avoid** in real scenarios. You can do this with your own project or you can download a public project named qmm-book-neighborhood-survey-collab. It already contains a conflict file.

1. Make sure you have the current (synchronized) version of the project both in QGIS and in Mergin Maps mobile app.

2. In QGIS, select an existing feature from a survey layer and modify its geometry.

3. In Mergin Maps mobile app and select the same feature. Use the ⬤ Edit button to open the form. Then tap [Edit geometry] and use the ⋯ button to select the *Redraw geometry* option. Modify the geometry (figure 6.12, on the next page) and save the changes [Record]. You can also change some attribute values in the form. Once you are finished, use the ✓ button to save the changes.

4. Synchronize the changes in the mobile app [Sync].

5. Now go back to QGIS and synchronize the changes ⟳. As changes made in QGIS were synchronized last, they will be displayed in the map window.

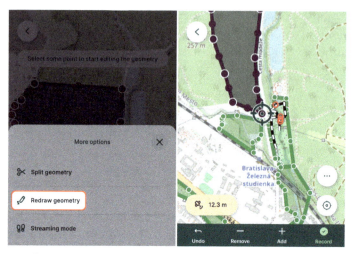

Figure 6.12: Redraw Geometry Option in Mergin Maps Mobile App

Changes made in the mobile app are considered to be conflicted by Mergin Maps.

6. A conflict file was added to the project. Go to the *Projects* tab in Mergin Maps dashboard (app.merginmaps.com). There will be a warning displayed next to the project name (figure 6.13) indicating there is a conflicting file. Open the project to see the conflict file (figure 6.12). Here it is a file named parks (edit conflict, sarah v36.json) meaning the conflict is in the parks layer, version v36, created by sarah. This version was later modified in a way that could not be resolved during the synchronization and some changes were rewritten. The conflict file can be opened and inspected for details.

My projects

Figure 6.13: A Project with a Conflict File in Mergin Maps Dashboard

7. Navigate to the *History* tab to see the last project versions (figure 6.15, on the facing page). Here we have the v36 version that contains the original data (before conflicting changes), v37 is the version created by alexandra and v38 is the current version of the project. The conflict file name indicates that some edits present in the version v37 were rewritten by the v38 version.

Figure 6.14: A Conflict File in Mergin Maps Dashboard

Figure 6.15: Project History in Mergin Maps Dashboard

8. In QGIS, use the *Create diff* processing algorithm (see Task 5 - Using Mergin Maps Processing Algorithms, on page 54) to compare project versions *v36* and *v37* (figure 6.15).

Figure 6.16: Inspecting Changes in QGIS Using the Create Diff Processing Algorithm

9. Decide which option is the correct one. If needed, manually adjust the attributes and geometry in the project in QGIS. The conflict file can be deleted now. You can delete it in the project folder on your computer or through the dashboard.

10. Don't forget to save and synchronize your changes.

Another approach how to inspect conflicting changes is to download the previous project version to your computer:

1. Navigate to the *History* tab of the Mergin Maps dashboard (figure 6.15, on the previous page).

2. Download the version that contains the overwritten edits. In our example, it is the version v37. If you want to compare changes with the original features, download also the version before conflicts occurred (here v36).

3. The project is downloaded as a zip file. Extract it and open the folder. As the conflict is in the *parks* layer, add the parks.gpkg to your project in QGIS and compare it with the current version (figure 6.17).

Figure 6.17: Inspecting Conflicting Geometry Changes in QGIS

6.5   Task 4 - Deploy a new project version

In the previous task we have dealt with a conflict file caused by two editors editing the same feature.

Synchronization issues can also occur when there are changes in the project and users do not use the same project version for their work. If you need to change the data schema or do significant changes in the project setup (add/delete/rename a field, change data types of the fields, setting up 1-N relations, etc.), it is for the best to follow these steps to avoid synchronization issues.

> Removing or adding existing fields to the form layout using Drag and drop designer or changing their aliases should not cause any issues. These changes can be synchronized as usual.

1. Synchronize all devices before making the changes in the project. Make sure there are no pending changes from any of the users who have access to the project.

2. The project needs to be removed from all devices. In QGIS, find the project in the *Browser panel* under the *Mergin Maps* provider. Right-click on it and *Remove locally* (figure 6.18). In Mergin Maps mobile app, tap the ⋮ button next to the project name and select the *Remove from device* option (figure 6.19, on the following page). The project is still saved in Mergin Maps cloud.

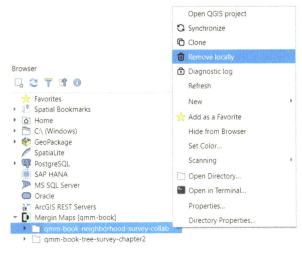

Figure 6.18: Removing a Project Locally through the Browser Panel in QGIS

3. There should be now only one local copy of the project (yours), so you can safely modify the data schema.

4. Synchronize the project to Mergin Maps. Your collaborators can now download the revised project and use it.

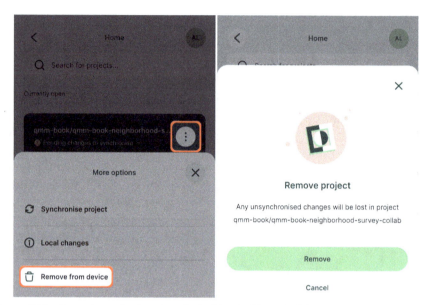

Figure 6.19: Removing a Project Locally in Mergin Maps Mobile App

7. Creating a Map

In this chapter you will learn the basics of composing a map of your field data in QGIS. You are encouraged to use a project with your own data. The example shown here is a version of the neighborhood survey project you cloned at the end of Chapter 3, but with features collected into the tree, path and park layers. If you need a project to work with, you can clone the public MM project named: `qmm-book-neighborhood-survey-chapter5`.

This exercise includes the following tasks:

- Task 1 - Layer Symbology.
- Task 2 - Setting up a Print Composition.
- Task 3 - Add and Configure Map Elements.
- Task 4 - Export the Map.

After this exercise you will be able to:

- Use the Categorical renderer.
- Use and configure SVG symbols.
- Compose a basic print layout.
- Add different map elements to a print composition.
- Configure map elements.
- Export a map.

7.1 Task 1 - Layer Symbology

To achieve a properly designed map, the features on the map must be easily distinguishable, attractive to the map reader, and stand out from the background data. Therefore, you may choose to adjust your layer styling from that used for data collection. If this is the case, it may be desirable to create a project file purely for cartographic output.

In this first task you will adjust layer symbology and learn to set up a print composition. You will then add and configure map elements such as a title, legend, north arrow, scale bar, and locator map. The final result will be a professional print composition.

Layer Symbology

1. The symbology set up for field data collection worked great for that purpose. However, here we will improve it for a map which will highlight the trees and path types in a park. Since data for paths and trees includes the path type and species of tree we will adjust the symbology to show the different types of each on the map. You will learn to create a copy of a layer for cartographic purposes. This will allow you to maintain the original layer for survey purposes.

2. Right-click on the paths layer and choose *Duplicate Layer* from the context menu.

3. Right-click on the duplicated layer and choose *Rename Layer*. Name this layer Paths by type.

4. Find the row of buttons atop the *Layers Panel*. To open the *Layer Styling Panel*, click the left most button which looks like a paint brush 🖌. You can also use the keyboard shortcut F7 to open this panel. The panel will open on the right side of the QGIS map canvas. Like all panels this can be undocked to a free-floating panel or moved to a different docking position. In a dual monitor environment, it can be nice to have the *Layer Styling Panel* in a different monitor. At the top of the panel is a drop-down menu that allows you to choose the target layer.

5. Set the target layer to Paths by type. Below the target layer is a dropdown for render type. Switch the renderer from *Single symbol* to *Categorized* renderer.

6. For *Value* choose the type and click the *Classify* button. QGIS will create a separate symbol for each individual value in the type field and one symbol for NULL values.

7. Since there are no NULL values select that category and click the *Delete* ⊟ button to remove it (figure 7.1, on the next page).

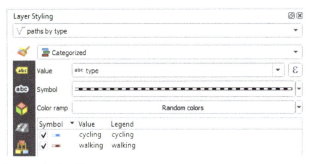

Figure 7.1: Styling Paths by Type

8. Next Duplicate the t reesurvey layer and name the copy Trees by type.

9. Turn your attention to the *Layer Styling Panel*. Set the target layer to Trees by type.

10. Select the *Simple marker* component.

11. Change the *Symbol layer type* to *SVG Marker*.

12. Scroll down in the *Layer Styling Panel* and find the *SVG Browser*. Use the search box beneath the *SVG Browser* to search for *tree* icons (figure 7.2).

Figure 7.2: SVG Browser

13. Select a tree icon and scroll back up to the settings. You can customize many aspects of an SVG symbol such as *Size, Fill color, Stroke color* and *Stroke width*. Use the example in figure 7.3, on the next page to make a nice tree symbol.

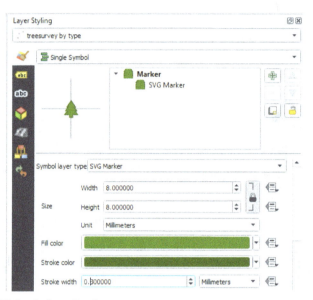

Figure 7.3: SVG Symbology Settings

ProTip: Notice that there are Data Defined Override icons 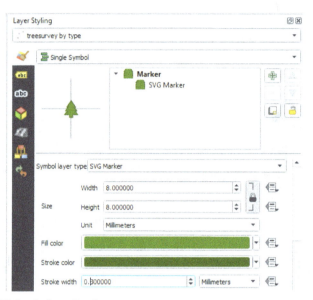 to the right of each symbol setting. These can be used to enhance the symbol by pointing to a field which contains values or by using an expression. To give the *Stroke color* a slightly darker color than the *Fill color* you can use the Darker() function with the @symbol_color variable. Left click on the Data Defined Override icon and choose *Edit* from the context menu. In the *Expression String Builder* search for the Darker() function. It is the Color group along with its companion function Lighter(). It just requires two inputs: a color and a factor. The @symbol_ color variable can be used for the color. Then enter a factor. Here 120 was used darker(@symbol_color ,120) to make the *Stroke color* slightly darker than the *Fill color*. Click *OK*. The Data Defined Override will now be yellow with an expression icon.

14. Now switch the renderer from *Single symbol* to *Categorized*.

15. For *Value* choose the tree_species and click the *Classify* button. QGIS will create a separate symbol for each individual value in the tree_species field and one symbol for NULL values. Delete the NULL class.

16. You can adjust the colors of each. If you used the Data Defined Override

the *Stroke color* will automatically adjust to be slightly darker than the *Fill color* for each symbol (figure 7.4).

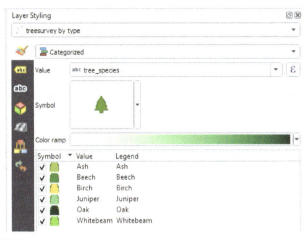

Figure 7.4: SVG Trees by Type

You now have some nicer symbology for your Neighborhood map.

7.2   Task 2 - Setting up a Print Composition

1. Next, you will open a New Print Layout. From the menu bar choose Project | New Print Layout, click the *New Print Layout* button or use the keyboard shortcut Ctrl + P.

2. Name the Layout with an appropriate name such as *Neighborhood Survey"* (figure 7.5).

Figure 7.5: Entering the Title of a New Print Layout

3. Click *OK*. A new Print Layout will open. This is where you craft your map.

The Print Layout is an application window with many tools that allow you to craft a map. The main window of the Print Layout displays the piece of paper upon which the map will be designed. There are buttons along the left side of the window that allow you to add various map elements: map, scale bar, photo, text, shapes, attribute tables, etc. (figure 7.6). Each item added to the map canvas becomes a graphic object that can be further manipulated (if selected) by the *Item Properties* tab on the right side of the layout. Across the top are buttons for exporting the composition, navigating within the composition and some other graphic tools (grouping/ungrouping etc.) For detailed information about the Print Layout, refer to the QGIS manual.[7]

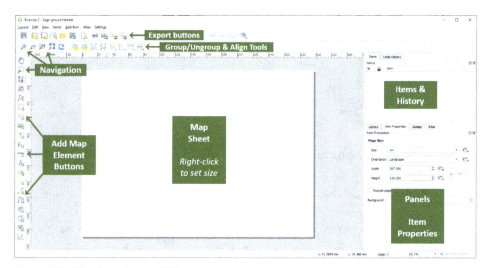

Figure 7.6: The Print Layout Window

4. Your first task will be to set the paper size. The default paper size is A4. You will keep this default. However, if you need to change the sheet size simply right-click on the blank page and choose *Page properties* from the context menu. The *Item Properties* tab changes to display *Page properties* (figure 7.7, on the next page).

> QGIS Print Layouts allow you to add as many pages to a Layout as you wish. These can also be of differing page sizes and orientations. You need to right-click on a given page to access the page dimensions.

---
[7]https://docs.qgis.org/3.28/en/docs/training_manual/map_composer/map_composer.html

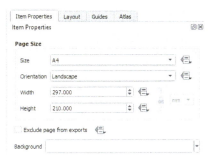

Figure 7.7: Print Layout Page Properties

5. Next set the *Orientation* to *Landscape*.

6. Using the *Add Map*  button, drag a box on the map sheet where you would like the map to go. Remember that you will need room for a title at the top of the page and a legend to the right of the map (figure 7.9, on page 119).

The map object can be resized after it has been added, by selecting it and using the handles around the perimeter to resize with the *Select/Move item* tool. This tool lets you select map features such as the map, a text block, legend etc. Remember when an object is selected, the *Item Properties* tab will show properties specific to that object.

7. With the map selected, click on the *Item Properties* tab. There is an embedded toolbar across the top of the panel (figure 7.8) with tools for controlling the map extent.

Figure 7.8: Map Layout Toolbar

From left to right these are:

- Update Map Preview
- Set Map Extent to Match Main Canvas Extent
- View Current Map Extent in Main Canvas
- Set Map Scale to Match Main Canvas Scale

- ![icon] Set Main Canvas to Match Current Map Scale

- ![icon] Bookmarks

- ![icon] Interactively Edit Map Extent

- ![icon] Labeling Settings

- ![icon] Clipping Settings

8. Click the *Set Map Extent to Match Main Canvas Extent* ![icon] button. This orients the map on the sheet of paper as it appears in the main QGIS map canvas.

9. To adjust the scale, change the *Scale* value. You will find it in the *Main properties* section. Map scale is a ratio of Map Distance/Ground Distance. Here the number is roughly 1000 which can be read as a scale of 1:1,000. To zoom out, increase this number. To zoom in, reduce this number. Clicking the *Refresh Map* ![icon] button forces the map view to refresh.

10. Scroll down and find the *Frame* setting and enable a frame for the map. Adjust the width as appropriate.

11. To pan the map, you can use the *Move Item Content* ![icon] button. This allows you to pan the map content in the map frame without changing the scale. It is normal to have to make adjustments to get the map extent just right. Try to make your map match the figure 7.9, on the facing page.

7.3 Task 3 - Add and Configure Map Elements

In this task you will learn how to add and configure the essential map elements. These include the title, legend, north arrow, scale bar and credits.

Adding the Title

The purpose of a map title is to quickly convey the content and focus of the map to the map reader. It should be concise and prominent.

1. Use the *Add Label* ![icon] tool to drag a box all the way across the top of the composition. The text box can be resized after the fact by using the graphic handles.

Figure 7.9: Print Layout Extent

2. By default, the text box will be populated with the placeholder text *Lorem ipsum*. Using the *Item Properties* tab *Main Properties* section, replace the holder text with the title: *'Neighborhood Survey'*.

3. In the *Appearance* section, find the large *Font* button and click on it to open the *Label Font* panel. Notice that there is the same series of text formatting tabs (*Formatting, Buffer, Mask & Background*), you find for labels in the main QGIS application (figure 7.10, on the next page). With these you can configure text buffers (halos), drop shadows, backgrounds, letter and word spacing etc., for all the text items in your layout. Set the *Font* to Times New Roman, *Size* 36 with a *Font style* of *Bold*. You may need to resize the text box after these font settings have been made so that the entire title can be seen. Click the blue *Go back* ◀ button to return to *Main Properties*.

> Clicking on the Settings | Layout Options menu will open the QGIS *Options* window to the *Layouts* tab. Here you can set the *Default* font for your layouts.

4. Note that you can also click the drop-down menu for *Font* and access a handy font widget (figure 7.11, on the following page). From there, you can interactively set the font size, pick from *Recent Fonts*, modify and copy the format, and use the color wheel/color patches to select a color.

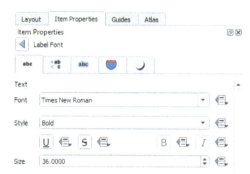

Figure 7.10: Labels Font Panel

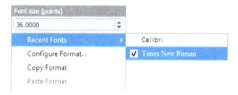

Figure 7.11: Upper Portion of the Layout Font Widget

5. Under *Appearance,* set *Horizontal alignment* to *Center* and *Vertical alignment* to *Middle.*

Setting Up the Legend

The purpose for the legend is to identify what different symbols and colors on the map represent. Legends are used for data layers that are non-intuitive or require more explanation. For example, a blue line labelled *Amazon River* is obviously a river. It does not need to be included in a legend. However, symbols for different types of paths do. Otherwise, the map reader will have no way of knowing what they are looking at.

1. Add a legend using the *Add new legend* button. Drag a box in an empty area of the map or in a blank space next to the map.

2. There may be layers that are only being used for cartographic purposes, for example base maps. These do not need to be in the legend. You will remove these unnecessary legend entries. On the *Item properties* tab, uncheck *Auto update.* Now the buttons below the *Legend Item* portion of the *Item properties*

tab are active.

From left to right these are:

- ▼ Move Item Down. Use to reorder layers & classes in the legend

- ▲ Move Item Up

- 🗐 Add Group

- 🕂 Add Item. Use to add a deleted or new layer to the legend

- ▭ Remove Item

- ✎ Edit Selected Item

- ε Add an Expression to the vector layer and each child symbol's label

- Σ Show feature count for each class of vector layer

- ε⌄ ▾ Filter Expression

3. Select each basemap and click the Delete ▭ button. Also remove the survey copy of the tree and path layers.

4. The Legend still needs some editing. Layers come into QGIS with the file name as the layer name. As you learned, you can rename layers in QGIS by right-clicking on the layer in the *Layers Panel* and choosing *Rename*. However, you can also edit these names directly in the Legend. Select the *treesurvey by type* legend entry and click the *Edit* ✎ button. The *Legend Item Properties* panel opens. *(Note: You can also double-click on a legend entry, to access the Legend Item Properties panel.)*

5. Rename the layer to *Trees*. Repeat this for *paths by type*, renaming it to simply *Paths* and rename *parks* with a capital P. Click the blue *Go back* ◀ button to return to *Main Properties*.

6. Expand *Fonts and Text Formatting*. Here you can configure fonts for *Item Labels, Subgroup Headings* etc. Set the *Item Labels* to *Calibri* with a *Font Size* of 10. Set the *Subgroup Headings* to *Calibri* with a *Font Size* of 10 and a *Font Style* of *Bold*.

7. Notice the *Columns* section. With this you can break a single column legend into multiple columns. This is not needed here, but is often needed for larger legends.

8. You can also give the Legend item a *Frame* as you did with the map.

> Notice the setting *Only show items inside linked map*. This is a handy way to eliminate layers and classes not visible on the map.

Your map should resemble figure 7.12.

Figure 7.12: Map with Title and Legend

Adding Descriptive Text

It good practice to include credits for both data sources and cartography on the map. It can also be helpful to include information such as the date.

1. You will enter some descriptive text that tells the map reader where the data was obtained, who the cartographer was, and the date created. This will be done using the *Add new label* tool, the same tool you used to add the title. Add the label below the legend. Add the following text in the *Main properties* window of the *Item Properties* tab.

- Data Sources: Open Street Map
- Cartographer: <your name>
- Date: <todays date>

2. You can manually enter the date. However, it is also possible to use an expression to create *Dynamic Text* for the date.

3. Select *Dynamic Text | Current Date | Day Month Year* (figure 7.13). An expression for the current date is entered. This will automatically update each time you open this print layout. This means it is no longer necessary to remember to update this every time you edit the map.

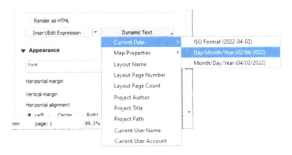

Figure 7.13: Dynamic Text Menu

4. Set the *Font* to a sans serif font such as Arial or Calibri. Set the *Font Size* to 8.

5. Position the text element in the lower right corner. With an element selected, you can also use the arrow keys on your keyboard to nudge it into place. Also notice that as you drag this map element composition guides appear. These are called *Smart Guides* and appear as an item aligns with the edges or centers of other map objects. You can use them to dynamically snap an item into place.

Adding a North Arrow

Often it is nice to add a north arrow to a map composition to help orient the map reader. One should especially be added if north is not up on the map. Here you will learn how to add this to your map.

> In the northern hemisphere map readers assume that north is up. On the Item Properties tab for the map object there is a Rotation section. This allows you to rotate the map. Sometimes rotating the map frame helps fit your study area onto the sheet of paper. If you have rotated your map, you should always include a north arrow.

1. Click on the *Add North Arrow* ![north arrow icon] button.

2. Drag a small box into the empty space in the lower right corner of the map.

3. A default north arrow is added. On the *Item Properties* the *SVG Browser* opens. Click the *Arrows* folder to find alternate north arrow styles. A series of SVG graphics included with QGIS will appear. Select a suitable icon.

4. Scroll down to the *Image Rotation* section of *Item Properties*. Here are the settings which orient the icon north. You can choose *Grid North* or *True North* (figure 7.14).

Figure 7.14: Aligning the North Arrow to Grid North

5. Scroll up to the *Size and Placement* section and change the *Resize mode* to *Zoom and resize frame*.

6. Resize the north arrow graphic and move as needed so that it is well placed.

Inserting a Scale Bar

Scale bars give the map reader a way to approximate distances on the map. There are three types: graphic scale bar, scale text and a verbal description, i.e.,

One cm equals 2 kilometers. Here you will learn how to add a graphic scale bar.

1. Click on the *Add new scalebar* button.

2. Drag a rectangle above the descriptive text to add the scalebar to the map.

3. In the *Main Properties* section click the drop-down menu for *Style*. There are eight different styles. Notice that one of the styles is *Numeric* which would add scale as scale text (*1:100,000*). Choose *Line Ticks Middle*.

4. On the *Item Properties* tab, in the *Units* section change the *Scalebar units* to *Meters*. This sets the scalebar units to meters.

5. In the *Segments* section, set the left to 1 and the right to 2.

6. Expand the *Display* section. This is where you can change the colors and width of the scale bar lines, as well as, the font. Click the *Font* button and change the *Font* Calibri *Size* 10.

7. Use the *Select/Move Item* tool to place the scalebar in a good position between the descriptive text and the legend. (figure 7.15, on the following page).

7.4 Task 4 - Exporting the Map

Congratulations your map is finished! The final step is to export it to a high-resolution pdf or jpg image.

1. Now that you have finished the map you will export it. The buttons below the *Settings* menu allow you to print the composition and export it into a variety of formats. These same options are available from the Layout menu.

2. Click the *Export as image* button. This will allow you to save the map out to an image file.

3. The *Save Layout As...* window opens. Navigate to the exercise folder. By clicking on the *Save as type* dropdown, you will see the range of image formats you can export to. Choose JPG. Name the file and click *Save*.

4. The *Image Export Options* window will open. It gives you the export settings including the pixel dimensions and the resolution you chose on the Composi-

Figure 7.15: Map with Scalebar, North Arrow, Legend and Text

tion tab of the Print Layout. Here you can change these setting if need be. Click *Save*. Note that it also has an option to *Generate a world file*!

5. You will receive the message that *Export layout: Successfully exported layout to...** (figure 7.16). This message includes a hyperlink to the folder where the map was exported. You can click on this to open your operating systems file browser and see the result.

Figure 7.16: Export Successful Notification with Hyperlink

6. Now click the *Export as PDF* button. Again, navigate to the exercise folder, name the file and click *Save*. During the PDF export you have the option of *Appending georeference information* and *Creating Geospatial PDF (GeoPDF)*. You do not need to do that here, but you may want to explore those settings.

7. Save you project and exit QGIS.

QGIS also has an *Atlas* feature.[a] You can use this to automate map production by creating a map series. To begin you can enable the *Atlas* panel by choosing *View - Panels - Atlas*. Here you can check *Generate an Atlas*, and configure the *Coverage Layer* and output file names. You can read the QGIS documentation to learn more.

---

[a]https://docs.qgis.org/3.28/en/docs/user_manual/print_composer/create_output.html#generate-an-atlas

8. Conclusions

In this book, we aimed to demonstrate the powerful combination of QGIS Desktop and Mergin Maps that makes it possible to take a QGIS project into field and seamlessly synchronize changes made by different users or on different devices.

Both QGIS and Mergin Maps are being actively developed. To keep up with these projects, check qgis.org or merginmaps.com respectively.

Although we tried to guide you through the task the best we could, maybe there will be some topics that you did not find explained as clearly as you would like. Also, you will for sure encounter some tasks that were not included in this book while working with Mergin Maps and QGIS. Luckily, both projects have online resources that can be helpful, such as:

- QGIS Desktop User Guide/Manual:
  https://docs.qgis.org/3.34/en/docs/user_manual/
- QGIS Training Manual:
  https://docs.qgis.org/3.34/en/docs/training_manual/
- Discover QGIS 3.x - Second Edition. A thorough treatment of all the features in QGIS authored by Kurt Menke and available from Locate Press: https://locatepress.com/book/dq32
- Mergin Maps documentation: https://merginmaps.com/docs/
- Hans van der Kwast YouTube account. Here you can find how-to videos on many aspects of QGIS and some on Mergin Maps: https://www.youtube.com/c/hansvanderkwast

Index

About Locate Press

Locate Press is a book publisher, focusing on the open source geospatial niche. Many traditional publishers see geospatial books as either scientific content or geared primarily toward consumers. Unfortunately, this means they don't give them the long term care they truly deserve. With more and more technical users using open source geospatial technology (for a wide variety of reasons!), now, more than ever, you need comprehensive and reliable education and training resources.

You've come to the right place!

We know that niche is not a swear word, but a marketplace that needs serious support. Geospatial data management is a core technology for government and business, making practical teaching materials for industry and higher education crucial. We also know that reliable availability of material is key. Our books, once available, remain available long after the first few thousand are sold so that you can depend on them for course material and reference long into the future.

If you are an educator looking for high quality curriculum, we would like to hear from you. Aside from training books, we also aim to provide workshop guides and exercise booklets that you can use in your courses!

Academia is not the only place for learning and training, so Locate Press supports consultants delivering workshops and seminars. If you have solid, practical material that needs some professional polish, give us a call. Likewise, if you need bulk orders to serve your students, or to resell, we can help there too.

Locate Press was founded by Tyler Mitchell in 2012, it's flagship book being *The Geospatial Desktop* by Gary Sherman. In 2013 Gary Sherman took over the helm as publisher, guiding the company until 2021 when Locate Press returned to Tyler.

ORDER DIRECT AND SAVE UP TO 30%

Our paperbacks can be ordered directly from us with bulk discounts on 5, 10, and 25+ units: store.locatepress.com

We print in countries that are closest to you and can deliver almost anywhere. Our print books also sell through Amazon or Ingram.

E-books (PDF) are ordered and downloaded directly from locatepress.com/ebooks

Educators contact us for desk/review copies:
+1 (250) 303-1831 or
tyler@locatepress.com

Writing for Locate Press

Are you passionate about open source software? Have an uncontrollable urge to share your knowledge with the world?

At Locate Press we're looking for books that open up the world of geospatial. We love concise, targeted titles that help people expand their knowledge and get up to speed quickly. That being said, we don't go around with blinders on—we're open to other leading edge topics related to open source.

We help put your ideas into book form, getting your expertise on paper and in print. Don't let the process scare you, we're here to guide and help all along the way—from outline to print-ready copy.

Le guide du programmeur PyQGIS
Gary Sherman, Noureddine Farah

Spatial SQL
Matthew Forrest

How to Succeed as a GIS Rebel
Mark Seibel

Earth Engine & Geemap
Qiusheng Wu

Using R as a GIS
Dr. Nick Bearman

Discover QGIS 3.x - 2nd Edition
Kurt Menke

QGIS for Hydrological Applications - 2nd Edition
Hans van der Kwast & Kurt Menke

Introduction to QGIS
Scott Madry Ph.D.

Leaflet Cookbook
Numa Gremling

QGIS Map Design - 2nd Edition
Anita Graser & Gretchen N. Peterson

The PyQGIS Programmer's Guide 3
Gary Sherman

pgRouting
Regina O. Obe & Leo S. Hsu

Geospatial Power Tools
Tyler Mitchell, GDAL Developers

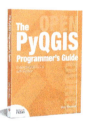

The PyQGIS Programmer's Guide
Gary Sherman

The Geospatial Desktop
Gary Sherman

locatepress.com

LOCATE PRESS
OPEN SOURCE
GEOSPATIAL | BOOKS

📞 +1 (250) 303-1831   ✉ tyler@locatepress.com

www.ingramcontent.com/pod-product-compliance
Lightning Source LLC
LaVergne TN
LVHW012331060326
832902LV00011B/1827